Tourism on the Verge

Series Editor

Ulrike Gretzel, Annenberg School for Communication and Journalism, University of Southern California, Los Angeles, CA, USA

"Tourism on the Verge" aims to provide a holistic understanding of various phenomena that shape tourism and hospitality in profound and lasting ways, approaching research topics and practical issues from multiple perspectives. Each volume in the series will address transformations within a particular area to advance both our theoretical understanding and practical applications. The books should be conceptual in nature and make highly relevant contributions to explaining these phenomena. Attention should also be given to cutting-edge methods to stimulate new directions in tourism research. The series will publish works of the highest quality that follow a logical structure, rather than merely presenting a collection of loosely related articles on a topic. Book editors will be asked to write a strong introductory chapter that offers a comprehensive overview of the selected topic areas/fields. Presenting a unique blend of scholarly research, the series will serve as an unparalleled reference source.

This is a SCOPUS indexed series.

Martina Pásková • Josef Zelenka

Social Responsibility in Tourism

Applications, Best-Practices, and Case Studies

Springer

Martina Pásková
University of Hradec Králové
Hradec Králové, Czech Republic

Josef Zelenka
University of Hradec Králové
Hradec Králové, Czech Republic

ISSN 2366-2611 ISSN 2366-262X (electronic)
Tourism on the Verge
ISBN 978-3-031-61608-2 ISBN 978-3-031-61609-9 (eBook)
https://doi.org/10.1007/978-3-031-61609-9

This Springer imprint is published by the registered company Springer Nature Switzerland AG
The registered company address is: Gewerbestrasse 11, 6330 Cham, Switzerland

Acknowledgment

The book was supported by the FIM UHK under the Grant of the Specific Research Project "Information and knowledge management and cognitive science in tourism". The authors thank FIM UHK student Zuzana Kroulíková for the innovative creation of diagrams in the text.

About the Book

This book explores the fundamental principles of social, cultural, environmental, and economic responsibility in tourism development, examining their relationship to sustainable tourism. It highlights the significance of local conditions in this context and discusses the challenges, opportunities, and problems and barriers associated with implementing responsible tourism. The aim is to guide tourism actors toward seeking balanced alternatives to current tourism development practices that can preserve or even improve conditions for future development.

Combining theoretical perspectives with practical insights, the book provides a comprehensive understanding of the principles, contexts, and circumstances surrounding responsible tourism approaches. It includes references to other sources of information, along with diagrams, and offers an overview and discussion of how social responsibility is applied by various tourism actors, including individuals, businesses, organizations, and destination management bodies. Concrete examples, case studies, and holistic approaches to addressing the environmental, economic, and socio-cultural impacts of tourism are given special emphasis.

This book is intended for anyone willing to acknowledge the shortcomings of current tourism development and understand the reasons behind its lack of sustainability. It aims to attract those interested in finding solutions to critical issues in tourism, revealing the potential and proven effects of tourism responsibility. In essence, the book primarily serves as a guide for tourism actors seeking to adopt more sustainable practices.

Contents

Chapter 1
Introduction to Topic

A consistent and globally implemented sustainable tourism has been a goal for nearly half a century, sought by tourism practitioners and experts from various scientific disciplines. However, as of 2024, when this book was completed, significant progress has been made in managing critical factors of tourism development (e.g. [1]) and conducting holistic research into sustainable tourism. This progress mainly stems from applying insights from disciplines like sociology, ecology, anthropology, philosophy, economy, geography, management and psychology. By taking this interdisciplinary approach, we gain a deeper understanding of tourism development conditions and factors, enhancing our ability to ensure sustainability.

Increasingly, sustainability in tourism is linked with promoting responsible behaviour amongst tourism actors (e.g. [2, 3, 4, 5]). This includes participants, managers, intermediaries, service providers and local and Indigenous peoples. Their growing level of responsibility is pivotal for practically implementing sustainable tourism ([6]).

The SARS-COV-2 pandemic offers a great opportunity for fundamental changes in tourism. Authors such as Cave, Dredge [7], Higgins-Desbiolles [8, 9], Rastegar, Higgins-Desbiolles, Ruhanen [10] and Talwar et al. [11] argue that this pandemic a global tourism crisis ([12]) prompts us to reevaluate tourism's benefits and adverse effects. It also encourages exploring alternatives like virtual reality tourism and online events, which gained prominence during the pandemic ([11]). These challenges have sparked broader efforts to reform tourism with "*wider efforts to reform tourism to be more ethical, responsible and sustainable*" ([9]:1).

The pandemic's lockdown measures provided unique conditions for evaluating tourism's impacts, highlighting both its benefits and its damages to destinations. This led to increased interest in slowing down tourism intensity and growth rates (Sharpley [13, 14]), advocating for concepts like "Degrowth in tourism" ([13, 14, 15, 16, 17]). This approach prioritises the rights of local residents over those of tourists and tourism entrepreneurs, emphasising travel closer to home. Additionally, the

M. Pásková, J. Zelenka, *Social Responsibility in Tourism*, Tourism on the Verge,
https://doi.org/10.1007/978-3-031-61609-9_1

pandemic underscored the importance of sustainable business practices for enhancing the tourism industry's resilience against similar crises ([18, 19, 20]).

Transitioning to low-carbon and potentially carbon-free processes in tourism is a fundamental challenge, given the accelerating impacts of climate change. Implementing social responsibility in tourism offers both global and local opportunities to mitigate and adapt to these effects ([21]).

Currently, the long-term impacts of tourism are socially unacceptable, economically unreasonable and environmentally unsustainable. The tourism economy generates significant externalities and self-destructive effects, resembling the tragedy of the commons. In this way, the potential for tourism development is destroyed by tourism itself. Unequal distribution of tourism's benefits and negative impacts leads to decreasing local acceptance of tourism. Moreover, the gradual destruction of nature and local landscapes is perceived negatively by both locals and visitors. The concept of "regenerative tourism" seeks to extend the benefits of tourism to the regeneration of local ecosystems, cultures and economies, requiring innovative approaches ([22, 23, 24]).

This publication explores the principles of sustainable tourism development, focusing on social, environmental and economic dimensions of responsible tourism. It emphasises the benefits, local specificities, shared values and appropriate approaches. The aim is to promote sustainable tourism development by fostering responsible behaviour amongst tourism actors. This includes grassroots actors like local and Indigenous peoples, as well as professionals such as destination managers and tourism service providers. These experts must understand tourism comprehensively, considering its impacts on a wider scale and over a longer timeframe.

The publication is based on the current state of knowledge about social responsibility in tourism. It brings original findings from the authors' research and aspires to interdisciplinary integration of this knowledge into a consistent system with an emphasis on applying theoretical concepts and practical examples from economics, geography, sociology, psychology, ethics, environmental studies, ecology and anthropology. It includes illustrative case studies, examples of good practice and an analysis of selected sustainable forms of tourism such as volunteer tourism, ecotourism and geotourism, which assume a minimum level of responsibility on both the supply and demand side of tourism. The term social responsibility is used to distinguish general social responsibility from the narrower meaning emphasising the social-cultural dimension of social responsibility (see Sect. 3.2.1).

Throughout this book, there is a strong emphasis on a proactive approach to responsibility taken by tourism operators, intermediaries, destination managers and participants. They aim to minimise or eliminate the negative impacts of their activities on society and the environment. The alternative approach, which involves compensating for negative impacts with projects that benefit society, should only be considered when it is impossible to avoid the negative impacts. Responsible action by tourism actors involves three dimensions: cognitive (learning about tourism's effects), ethical (incorporating this knowledge into decision-making) and relational/emotional (developing personal connections to affected people and places).

In the realm of responsible tourism implementation, tourism actors define their roles concerning natural and cultural heritage, local communities, other tourism actors and society. For example, when proposing tourism development focused on a destination's environmental or socio-cultural qualities, it is crucial to evaluate potential negative impacts such as reduced quality or unwanted changes in the area's character. Tourism stakeholders must acknowledge their role in these changes and weigh the undesirable impacts against expected benefits, discussing these values with experts and the affected public.

From the tourist's perspective according to Zelenka and Pásková [25], responsibility is evident in various choices, including destination selection (e.g. [26]), transportation modes (Fig. 6.6), accommodation preferences (e.g. favouring locally run facilities), tourism activities (e.g. opting for low-impact forms like ecotourism and geotourism), and engagement with local culture and customs. Developing basic language skills and using them to interact with the local community and create an unmediated relationship with them ([27]), choosing gifts and food to buy (preference for local cuisine and souvenirs) and guides (preference for locals) is also important. Responsibilities of tourism participants are discussed in detail in Sect. 5.6.

In the case of providers and intermediaries of tourism services and destination management, this responsibility is mainly manifested in:

- Respect for the local community, its needs and values, lifestyle, culture, customs, traditions etc.
- Determining the degree and manner of local community involvement in the provision of tourism services, tourist attractions with a preference for local community participation and in tourism products. Positive confirmation of the possibility of local community participation and the conditions for such participation are reported by Hamilton and Alexander [28]. Deery, Jago and Fredline [29], amongst others, state that local people and the life of local communities are an important part of tourism products, local people need to be allowed to participate in the decision-making process, and the impacts of tourism on them should be closely monitored.
- Determining the intensity and forms of tourism development, including the level and type of tourism infrastructure development.
- Deciding the extent and manner of local communities' inclusion in the offer of tourism products.
- Reducing the level of negative impacts of their activities, in particular, by implementing EMS (Environmental Management System), green management and consistent waste management, as described in detail in Sects. 5.1, 5.2 and 5.3.
- Compensating local people, communities and the biosphere in general for the unavoidable environmental impacts of tourism on the destination (Sect. 3.5).

In addition to social responsibility for the consequences of the provision of tourism services and regional responsibility, the responsibility of tourism for reducing global problems is also declared and implemented. This responsibility focuses, in particular, on the problems of global warming, tropical rainforest area reduction, xenophobia and tensions in the world, and the protection and conservation of geodiversity, cultural diversity and biodiversity (on the application of social responsibility in relation to biodiversity conservation, see Marsden et al. [30]). This global responsibility is identified and declared (e.g. the UN WTO Global Code of Ethics for Tourism) and, above all, implemented through practical activities. These include the introduction of green management, airline carbon offset programmes and a preference for the development of appropriate forms of tourism. In a number of studies (e.g. [31, 32, 33]), the extent to which ecotourism contributes to the protection of biodiversity, including the protection of tropical rainforests, is specifically discussed with regard to the way social responsibility is implemented.

The role of guides, animators and interpreters, who directly interact with visitors, is crucial in shaping visitors' perceptions and acceptance of local culture. At the same time, they also have a significant influence on the willingness of visitors to accept local culture and on their possible enrichment with local specificities and culture, including "other dimensions" of reality (Fig. 1.1). The ability to accept the specificities of another culture, and even to enrich one's mental and emotional world with them, increases the level of respect and responsibility for the consequences of one's own behaviour.

As public awareness of environmental consequences grows, methods like carbon and water footprinting are increasingly used to express social and environmental responsibility ([34]). The carbon footprint for calculating the impact of tourism

Fig. 1.1 The reception area of a small hotel in the Amazon reflects the cultural specificities of the region—the physical and spiritual connection to luxuriant vegetation, the love of colour and the creativity unbridled by the aesthetic norms of the Western world (Napo Sumaco aspiring UNESCO Global Geopark, Ecuador). Photo: M. Pásková, 2022

investments was described by Cadarso et al. [35]). Resource consumption by tourism was discussed by Munday, Turner and Jones [36]. Gössling and Buckley [37] also addressed the limits in the functioning of carbon footprint-based eco-labelling, which has so far been plagued by ineffective communication with actual and potential customers.

Personal or group social responsibility can also rely on a more intricate measure called the ecological footprint, which gauges the overall impact of a specific activity on the Earth ([38]). Kettl [39] applied an ecological footprint-based method to compare the environmental impacts of different transportation modes. Additionally, Marzouki, Froger and Ballet [40] conducted a comparison of the environmental impacts of ecotourism and mass tourism using ecological footprint analysis. Michailidou et al. [41] utilised a combination of ecological footprint analysis and life cycle assessment methods to analyse the impact of tourism on the Greek peninsula of Chalkidiki. Sun [42] analysed the impacts of tourism infrastructure, transportation of tourists by aircraft and importation of supplies for visitors to small islands using the ecological footprint. Various types of footprints, including environmental, social, economic, combined and composite, are utilised based on the fundamental dimensions of sustainability ([43]), with methods provided for their calculation.

Agenda 21 can be regarded as a significant declaration of humanity's responsibility towards nature and the advancement of our species. It acknowledges our capacity for progress as well as our potential to cause harm or protect the environment to ensure the survival of all plant and animal species. Agreed upon at the Earth Summit in Rio de Janeiro in 1992, Agenda 21 laid out the theoretical definition and declaration of sustainable development and gradually put into practice using methodologies such as those developed by UN WTO [44], UNEP [45], Vourc'h, Denman [46] as well as tourism policies at different geographical levels of tourism development and through tourism policies at various geographical levels of development.

The general concept of social responsibility is also expanding in terms of quality, content and understanding of interrelationships in several different directions—in terms of its content, geographical scope and whether there is only one entity claiming social responsibility, or a concerted involvement of various entities, individuals and administrations at the local, regional or global level. This broadening of the concept of social responsibility geographically, in terms of content, system and in terms of cooperation, as well as the degree of involvement of other tourism stakeholders and actors, is illustrated clearly in Fig. 1.2.

In addition to research, declaration and implementation of social responsibility in large organisations, there is a growing focus on social responsibility of:

- **Small and medium-sized businesses**—for firms in the tourism industry, see, e.g. Tamajón and Font [11], examples from other sectors are exemplified by Koubská and Hralová [47].
- **Individuals**—visitor responsibility in community tourism is discussed by Sin and Minca [48].

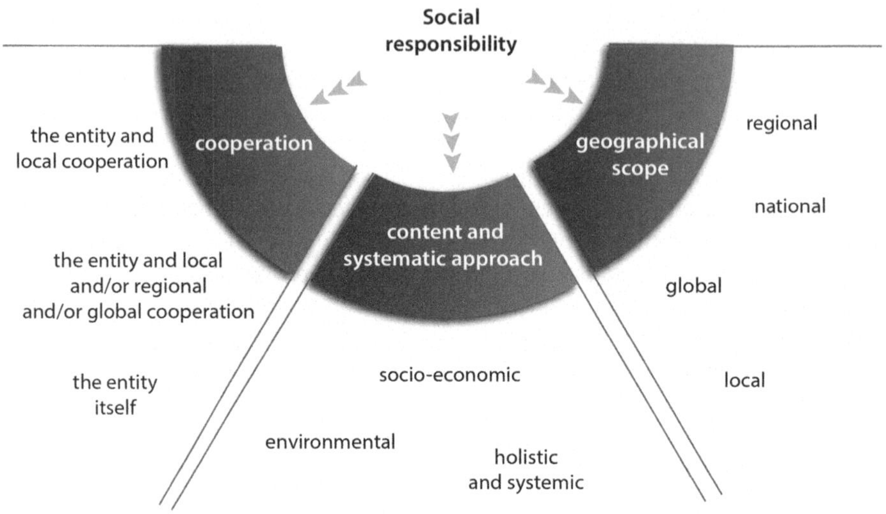

Fig. 1.2 Expanding the concept of social responsibility in three different contexts—geographical, content, systematic and cooperative

Fig. 1.3 The interplay between the concept of sustainable tourism, which addresses the impacts of tourism, and the concept of social responsibility in tourism, which addresses the attitudes and behaviour of tourism actors

- **Local community**—Su, Wang and Wen [49] present local community approaches to tourism development.
- **Other tourism actors and stakeholders**.

Simply put, the concept of sustainability in tourism focuses on the effects of tourism, their optimisation and the appropriate procedures for achieving this optimisation, whilst the concept of social responsibility focuses on the approach of individual tourism actors and tourism organisations to its implementation (Fig. 1.3). Thus, social responsibility in tourism includes the approaches of individual tourism actors and tourism organisations means not only to prevent unsustainable tourism impacts, but also to realise positive externalities. Socially, environmentally and economically responsible development of tourism constitutes the content of the concept of sustainable development of tourism, as discussed in Chap. 2.

The following chapters offer answers to the following questions in the context of sustainable development and social responsibility in tourism:

- What is social responsibility in tourism, and how is it changing over time (see, in particular, Sects. 3.1 and 3.2)?
- What motivates different people in the tourism industry to adopt social responsibility or take part in it?
- Why the current mode and intensity of tourism is no longer sustainable?
- What does sustainable tourism mean, and how does it relate to social responsibility (see, in particular, Chap. 2)?
- What makes social responsibility in tourism unique?
- How do declarations, charters, codes of ethics and international standards play a role in social responsibility in tourism (see, in particular, Sects. 4.4, 4.5, 4.6, 4.7 and 4.8)?
- How can we see social responsibility in the actions of tourism service providers, destination managers, tourists, and others involved in tourism?
- Which types of tourism are currently most relevant for applying the concept of social responsibility (see Chap. 6)?

References

1. Henderson, J. C. (2007b). *Tourism crises: Causes, consequences and management* (p. 186). Routledge.
2. Caruana, R., et al. (2014). Tourists' accounts of responsible tourism. *Annals of Tourism Research, 46*, 115–129. https://doi.org/10.1016/j.annals.2014.03.006
3. Charter. (2022). The Moroccan charter of responsible tourism, [online]. Retrieved August 04, 2022, from https://www.iamorocco.com/en/portal-of-morocco/tourism/responsible-tourism/
4. Ethical Travel Portal. (2022). Ethical travel portal, How to be a responsible tourist? [online]. Retrieved October 4, 2022, from https://ethicaltravelportal.com/, http://ethicaltravelportal.com/how-to-be-a-responsible-traveller/
5. Medina, L. K. (2005). Ecotourism and certification: Confronting the principles and pragmatics of socially responsible tourism. *Journal of Sustainable Tourism, 13*(3), 281–295. https://doi.org/10.1080/01434630508668557
6. Boluk, K. A., Cavaliere, C. T., & Higgins-Desbiolles, F. (2019). A critical framework for interrogating the United Nations Sustainable Development Goals 2030 Agenda in tourism. *Journal of Sustainable Tourism, 27*(7), 847–864. https://doi.org/10.1080/09669582.2019.1619748
7. Cave, J., & Dredge, D. (2020). Regenerative tourism needs diverse economic practices. *Tourism Geographies, 22*(3), 503–513. https://doi.org/10.1080/14616688.2020.1768434
8. Higgins-Desbiolles, F. (2020). Socialising tourism for social and ecological justice after COVID-19. *Tourism Geographies, 22*(3), 610–623. https://doi.org/10.1080/14616688.2020.1757748

9. Higgins-Desbiolles, F. (2021). The "war over tourism": Challenges to sustainable tourism in the tourism academy after COVID-19. *Journal of Sustainable Tourism, 29*(4), 551–569. https://doi.org/10.1080/09669582.2020.1803334
10. Quazi, H. A., Khoo, Y.-K., Tan, C.-M., & Wong, P.-S. (2001). Motivation for ISO 14000 certification: Development of a predictive model. *Omega, 29*(6), 525–542. https://doi.org/10.1016/S0305-0483(01)00042-1
11. Stem, C. J., et al. (2003). How 'eco' is ecotourism? A comparative case study of ecotourism in Costa Rica. *Journal of Sustainable Tourism, 11*(4), 322–347. https://doi.org/10.1080/09669580308667210
12. Cheer, J. M., Lapointe, D., Mostafanezhad, M., & Jamal, T. (2021). Global tourism in crisis: Conceptual frameworks for research and practice. *Journal of Tourism Futures, 7*(3), 278–294. https://doi.org/10.1108/JTF-09-2021-227
13. Schianetz, K., Kavanagh, L., & Lockington, D. (2007). The learning tourism destination: The potential of a learning organisation approach for improving the sustainability of tourism destinations. *Tourism Management, 28*(6), 1485–1496. https://doi.org/10.1016/j.tourman.2007.01.012
14. Schwarz, W., & Schwarz, D. (1998). *Living lightly. Travels in post-consumer society* (p. 400). John Carpenter Publishing.
15. Fletcher, R., Mas, I. M., Blanco-Romero, A., & Blázquez-Salom, M. (2019). Tourism and degrowth: An emerging agenda for research and praxis. *Journal of Sustainable Tourism, 27*(12), 1745–1763. https://doi.org/10.1080/09669582.2019.1679822
16. Higgins-Desbiolles, F. (2022). Subsidiarity in tourism and travel circuits in the face of climate crisis. *Current Issues in Tourism, 26*(19), 3091–3310. https://doi.org/10.1080/13683500.2022.2116306
17. Higgins-Desbiolles, F., Carnicelli, S., Krolikowski, C., Wijesinghe, G., & Boluk, K. (2019). Degrowing tourism: Rethinking tourism. *Journal of Sustainable Tourism, 27*(12), 1926–1944. https://doi.org/10.1080/09669582.2019.1601732
18. Dogru, T., Akyildirim, E., Cepni, O., Ozdemir, O., Sharma, A., & Yilmaz, M. H. (2022). The effect of environmental, social and governance risks. *Annals of Tourism Research., 95*, 103432. https://doi.org/10.1016/j.annals.2022.103432
19. Klein, R. A. (2011). Responsible cruise tourism: Issues of cruise tourism and sustainability. *Journal of Hospitality and Tourism Management, 18*(1), 107–116. https://doi.org/10.1375/jhtm.18.1.107
20. Pizzitutti, F., Mena, C. F., & Walsh, S. J. (2014). Modelling tourism in the Galapagos Island: An agent-based model approach. *Journal of Artificial Societies and Social Simulation, 17*(1), 14. https://doi.org/10.18564/jasss.2389
21. Popa, R. A. (2015). The corporate social responsibility practices in the context of sustainable development. The case of Romania. *Procedia Economics and Finance, 23*, 1279–1285. https://doi.org/10.1016/S2212-5671(15)00395-0
22. Bellato, L., Frantzeskaki, N., Lee, E., Cheer, J. M., & Peters, A. (2023). Transformative epistemologies for regenerative tourism: Towards a decolonial paradigm in science and practice? *Journal of Sustainable Tourism, 32*(9), 1–21. https://doi.org/10.1080/09669582.2023.2208310
23. Bellato, L., Frantzeskaki, N., & Nygaard, C. A. (2023). Regenerative tourism: A conceptual framework leveraging theory and practice. *Tourism Geographies, 25*(4), 1026–1046. https://doi.org/10.1080/14616688.2022.2044376
24. Dredge, D. (2022). Regenerative tourism: Transforming mindsets, systems and practices. *Journal of Tourism Futures, 8*(3), 269–281. https://doi.org/10.1108/JTF-01-2022-0015
25. Zambrano, A. M. A., Broadbent, E. N., & Durham, W. H. (2010). Social and environmental effects of ecotourism in the Osa Peninsula of Costa Rica: the Lapa Rios case. *Journal of Ecotourism, 9*(1), 62–83. https://doi.org/10.1080/14724040902953076
26. Sörensson, A., & von Friedrichs, Y. (2013). An importance–performance analysis of sustainable tourism: A comparison between international and national tourists. *Journal of Destination Marketing & Management, 2*(1), 14–21. https://doi.org/10.1016/j.jdmm.2012.11.002

27. Jafari, J. (Ed.). (2000). *Encyclopedia of tourism* (p. 683). Routledge. ISBN 0415154057.
28. Hamilton, K., & Alexander, M. (2013). Organic community tourism: A cocreated approach. *Annals of Tourism Research, 42*, 169–190. https://doi.org/10.1016/j.annals.2013.01.015
29. Deery, M., Jago, L., & Fredline, L. (2012). Rethinking social impacts of tourism research: A new research agenda. *Tourism Management, 33*(1), 64–73. https://doi.org/10.1016/j.tourman.2011.01.026
30. Ma, J. (2012). A study on the models for corporate social responsibility of small and medium enterprises. *Physics Procedia, 25*, 435–442. https://doi.org/10.1016/j.phpro.2012.03.108
31. Kilipiris, F., & Zardava, S. (2012). Developing sustainable tourism in a changing environment: Issues for the tourism enterprises (Travel Agencies and Hospitality Enterprises). *Procedia – Social and Behavioral Sciences, 44*, 44–52. https://doi.org/10.1016/j.sbspro.2012.05.003
32. Lee, Y.-K., Choi, J., Moon, B., & Babin, B. J. (2014). Codes of ethics, corporate philanthropy, and employee responses. *International Journal of Hospitality Management, 39*, 97–106. https://doi.org/10.1016/j.ijhm.2014.02.005
33. Xu, J., Yao, L., & Mo, L. (2011). Simulation of low-carbon tourism in world natural and cultural heritage areas: An application to Shizhong District of Leshan City in China. *Energy Policy, 39*(7), 4298–4307. https://doi.org/10.1016/j.enpol.2011.04.047
34. Niehm, L. S., Swinney, J., & Miller, N. J. (2008). Community social responsibility and its consequences for family business performance. *Journal of Small Business Management, 46*(3), 331–350. https://doi.org/10.1111/j.1540-627X.2008.00247.x
35. Cadarso, M. A., Gómez, N., López, L. A., & Tobarra, M. Á. (2016). Calculating tourism's carbon footprint: Measuring the impact of investments. *Journal of Cleaner Production, 111*(Part B), 529–537. https://doi.org/10.1016/j.jclepro.2014.09.019
36. Mody, M., et al. (2014). The different shades of responsibility: Examining domestic and international travelers' motivations for responsible tourism in India. *Tourism Management Perspectives, 12*, 113–124. https://doi.org/10.1016/j.tmp.2014.09.008
37. Gössling, S., & Buckley, R. (2016). Carbon labels in tourism: Persuasive communication? *Journal of Cleaner Production, 111*(Part B), 358–369. https://doi.org/10.1016/j.jclepro.2014.08.067
38. Hunter, C., & Shaw, J. (2007). The ecological footprint as a key indicator of sustainable tourism. *Tourism Management, 28*(1), 46–57. https://doi.org/10.1016/j.tourman.2005.07.016
39. Kajala, L., et al. (2004). Measures for developing sustainability of nature tourism in protected areas. In Proceedings of the second international conference on monitoring and management of visitor flows in recreational and protected areas, June 16–20, Rovaniemi, Finland, p. 417, 236–241. ISBN 951-40-1930-X (PDF), [online]. Retrieved June 24, 2013, from http://www.metla.fi/julkaisut/workingpapers/2004/mwp002.htm
40. Mao, Z., Yang, Y., & Zhou, J. (2022). Are GreenLeaders also performance leaders? An econometric analysis of TripAdvisor hotel certification of GreenLeaders. *Journal of Sustainable Tourism, 31*(9), 2150–2172. https://doi.org/10.1080/09669582.2022.2127740
41. McCool, S. F., & Lime, D. W. (2001). Tourism carrying capacity: Tempting fantasy or useful reality. *Journal of Sustainable Tourism, 9*(5), 372–388. https://doi.org/10.1080/09669580108667409
42. Starwood. (2016). Starwood Hotels & Resorts Worldwide Inc. – Supplier code of conduct, [online]. Retrieved April 10, 2016, from https://www.starwoodhotels.com/Media/PDF/Corporate/SC_Supplier_Letter_-_Code_of_Conduct_Jan15_Final.pdf
43. Čuček, L., Klemeš, J. J., & Kravanja, Z. (2012). A review of footprint analysis tools for monitoring impacts on sustainability. *Journal of Cleaner Production, 34*, 9–20. https://doi.org/10.1016/j.jclepro.2012.02.036
44. Tsai, W. H., Chang, Y.-C., Lin, S.-J., Chen, H.-C., & Chu, P.-Y. (2014). A green approach to the weight reduction of aircraft cabins. *Journal of Air Transport Management, 40*, 65–77. https://doi.org/10.1016/j.jairtraman.2014.06.004

45. Tourisme Québec. (2017). The ecotourist code of conduct, [online]. Retrieved June 22, 2017, from https://www.tourisme.gouv.qc.ca/publications/media/document/etudes-statistiques/code_eco_a.pdf
46. UNWTO. (2022a). Sustainable Tourism Development. UN World Tourism Organization (UNWTO), [online]. Retrieved September 4, 2022, from https://www.unwto.org/sustainable-development
47. Kim, Y., & Han, H. (2010). Intention to pay conventional-hotel prices at a green hotel – A modification of the theory of planned behavior. *Journal of Sustainable Tourism, 18*(8), 997–1014. https://doi.org/10.1080/09669582.2010.490300
48. Sharpley, R. (2022). Tourism and development theory: Which way now? *Tourism Planning & Development, 19*(1), 1–12. https://doi.org/10.1080/21568316.2021.2021475
49. Stankey, G. H. (1981). Integrating wildland recreation research into decision making: Pitfalls and promises. *Recreation Research Review, 9*(1), 31–37.

Chapter 2
Concept and Context of Tourism Sustainability

2.1 The Historical Context of the Development of Tourism and the Introduction of More Sustainable Approaches to Its Development

Modern tourism, tracing its roots back to the mid-nineteenth century, underwent significant transformations until the 1970s. During this period, it was closely linked with the advancement of transportation, characterised by largely unregulated expansive growth and insufficient monitoring and assessment of its impacts and consequences. Mass tourism, characterised by relatively homogeneous offerings and standardised services, prevailed, prioritising short-term gains and economies of scale, particularly focusing on the economic dimensions of tourism. Examples such as hotel chains, colossal cruise ships and expansive seaside resorts exemplify the pursuit of economies of scale in mass tourism development. Consequently, tourism was often referred to as the "tourism industry", and destinations were developed akin to industrial zones, catering primarily to mass tourism.

Since the 1970s, the impacts of tourism on both natural and socio-cultural environments in heavily visited regions have garnered increasing attention from various academic disciplines including geography, economics, sociology, anthropology and psychology. During this period, several seminal publications emerged on this topic, including "Towards a Sociology of International Tourism" by Cohen [1], "Blessing or Blight?" by Young [2], "Why Destinations Rise and Fall in Popularity" by Plog [3], "The Golden Hordes" by Turner and Ash [4], "Second Homes: Curse or Blessing" by Coppock [5] and "Tourism: The Good, The Bad, and the Ugly" by Rosenow and Pulsipher [6].

Since then, the discourse surrounding the sustainability of tourism has evolved and become more refined, driven by the growing interest in interdisciplinary research. Various strands of inquiry have emerged, focusing on different aspects of tourism development. This includes research on the patterns of destination

M. Pásková, J. Zelenka, *Social Responsibility in Tourism*, Tourism on the Verge,
https://doi.org/10.1007/978-3-031-61609-9_2

development ([7, 8, 9, 10, 11, 12, 13, 14, 15, 16, 17, 18, 19]), the dynamics between host and visiting communities through the lens of social exchange theory ([1, 20, 21, 22, 23, 24, 25, 26, 27, 28, 29]), assessing the carrying capacity of regions for tourism development ([17, 18, 25, 26, 30, 31, 32]), and the phenomenon of overtourism, which refers to the perceived excessive pressure of tourism on local communities (e.g. [29, 32]).

Also, in the context of regional development, until the 1990s, the main emphasis was put on its economic and social aspects. The latter were mainly linked to the short-term economic status of the local community and employment in the tourism industry. From the perspective of the concept of development and vision of tourism, the period of 1980s and 1990s was a turning point, first in the developed countries and later worldwide, when the concept of sustainable development, and its key document—Agenda 21 for the Travel and Tourism Industry, was developed and gradually shaped. This concept has been gradually transferred to the tourism sector as part of concepts of tourism sustainability, based on earlier theoretical work aimed at clarifying the nature and key manifestations and mechanisms of tourism (Fig. 2.1), and is being further developed. Current directions for future research in sustainable tourism are outlined, for example by Buckley [34]. Due to the dynamics of the tourism industry, the concept of sustainable development is being applied in tourism quite rapidly, albeit so far rather marginally in terms of the overall scale of the

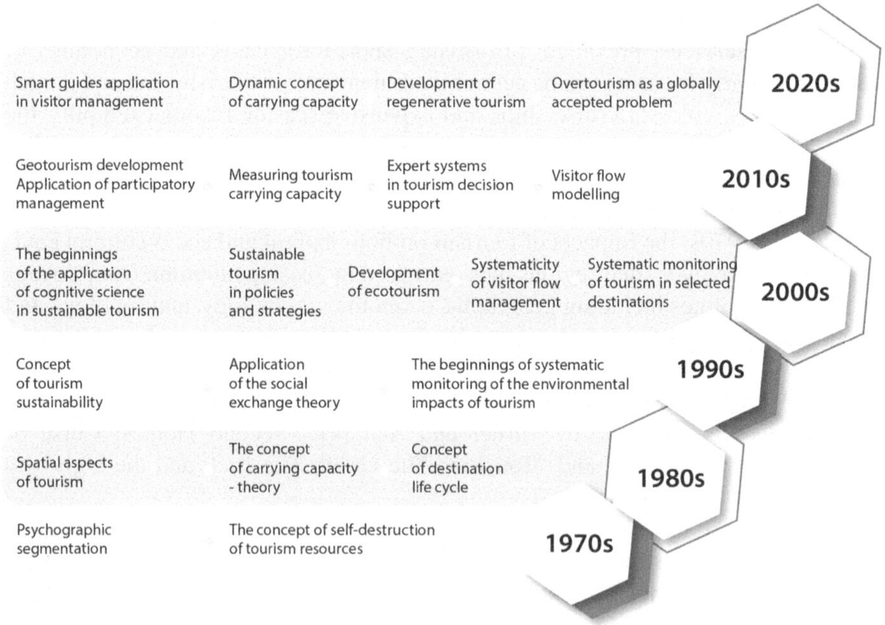

Fig. 2.1 Staging of identifying or introducing basic concepts and approaches to tourism sustainability. Source: [33]:13, supplemented

tourism industry. Change in the clients' mindsets, value systems, preferences,[1] mental maps, motivations, and behaviour patterns[2] reflect the abovementioned dynamic situation in the tourism sector. Similar changes can be observed in the mindsets of tourism entrepreneurs and other tourism actors.

The progressively developed concept of tourism sustainability is much more complex than mass tourism, which is not very structured. The concept of tourism sustainability seeks to balance specific local conditions and local potential for tourism development, with an emphasis on respect for, preservation and responsible use of these local specificities. The basic premise of tourism sustainability is to step out of the vicious circle (circus vitiosus; [37]), i.e. to overcome the "tourist trap effect" ([38, 39]), according to which the long-term unbearable burden caused mainly by excessive quantity and inadequate quality of secondary resources (infrastructure built for visitors) irreversibly damages the primary resources of the destination (cultural and natural heritage and pretouristic infrastructure). The relationship between tourism and natural resources, which they refer to as the "environmental paradox", is discussed by, for example Williams and Ponsford [40], and Pásková [41]. This is also the basis for the definition of sustainable tourism ([39, 41]) as tourism that does not irreversibly damage the natural, cultural and social environments even in the long-term perspective of its implementation (i.e. pin particular primary destination resources). Zelenka, Pásková and Husáková [42]:19–23 consider it appropriate to define sustainable tourism also in terms of processes and phenomena (i.e. to focus on concepts, goals, strategies and measures to be used to achieve sustainability, see also [43].

Another crucial aspect of considering tourism sustainability involves striving for a balance amongst different natural, social, and regional priorities. This entails the application of strategic planning in tourism ([38, 40, 44]), conducting research on tourism ([45]), and adopting a systemic approach to tourism research, education, and theory implementation ([46, 47, 48]). Numerous studies, such as the process-system analysis of tourism sustainability ([38, 49]), are grounded in the overarching concept of sustainable development. To ensure the sustainability of activities undertaken by various tourism stakeholders, specific recommendations are made:

- The importance of ensuring a balance, or even synergy, of the three dimensions of tourism sustainability, also referred to as the pillars of social development—economic, socio-cultural and environmental.
- Prevention of risks associated with tourism development—e.g. strengthening those components of the institutional environment that reduce the number or impact of negative market externalities.
- Exploiting opportunities—such as increasing civic engagement and responsibility, environmental innovations (reducing energy consumption, noise pollution,

[1] For example, research on wilderness visitor preferences in Ballantyne, Packer and Hughes [35] and visitor ecological preferences in Kelly et al. [36].

[2] For example, changing behaviour and mindsets of wilderness visitors in Ballantyne, Packer, and Hughes [35].

light pollution and other noxious pollutants etc.) in tourism and transport infrastructure.
- The need to optimise the effects of tourism ([38, 39, 41, 50, 51]) so that a compromise is always sought between maximising the benefits and minimising the adverse effects of tourism.

In the same vein, Romão et al. [52] cite the combination of nature and landscape conservation (i.e. the environmental dimension), economic prosperity and social equity, as well as benefits to the local community as potential opportunities for sustainable tourism. Some authors add another dimension to the three above, variously labelled as either institutional (e.g. [53, 54]) or spatial/territorial/regional ([55]) or cultural ([56]). Thereby, they demonstrate that tourism sustainability has more significant aspects and is conditioned by respect for local specificities (regional pillar), a mutually beneficial exchange between the guest (visitors) and host (local residents) communities and a quality institutional environment (local administration, legal frame, destination management).

The time factor is another important dimension of tourism sustainability that complements strategic planning, tourism impact management and the synergy of tourism and societal development dimensions. Indeed, the time aspect is one of the decisive factors for tourism sustainability. Typically, long-term character is assessed in relation to the partial or total effect of tourism. The degree of tourism seasonality (kurtosis) is important in terms of its economic profitability and the character of its impact on the natural and socio-cultural environment. The types of visitors to a destination, the way tourism infrastructure is developed and the rate of return on investment, e.g. in the greening of tourism infrastructure and staff training, also change over time.

The time factor is also important in tourism sustainability (as it is in environmentalism) in terms of whether the level of expected benefits can be achieved before the destination environment is damaged unacceptably. Tourism effects over time therefore have to be monitored. On the basis of the data obtained, it is possible to implement an anti-cyclical policy leading to the prevention of undesirable life cycle of the destination. In relation to the economic, socio-cultural and environmental pillars, qualitative indicators of sustainable tourism are most often cited as a complement and elaboration of this concept, as shown in the diagram in Fig. 2.2 and described in detail in Yunis [58], Janusz and Bajdor [43] and UNWTO [59]:18. The diagram also illustrates the most important aspects of the quality of the destination environment for which various tourism actors should take a sufficient degree of social responsibility.

According to Ali and Frew [60]:85–101, an alternative pragmatic approach to achieving tourism sustainability is the creation of a low-resource consumption tourism supply chain of suppliers, integrators and service providers. Low-resource consumption would be ensured by the use of Information and Communication Technology (ICT) in addition to the systemic approach of environmental management (see Sect. 4.8). An analysis of the implementation of such chains in tourism is presented by Marimon, Llach and Bernardo [61] and the creation of such networks

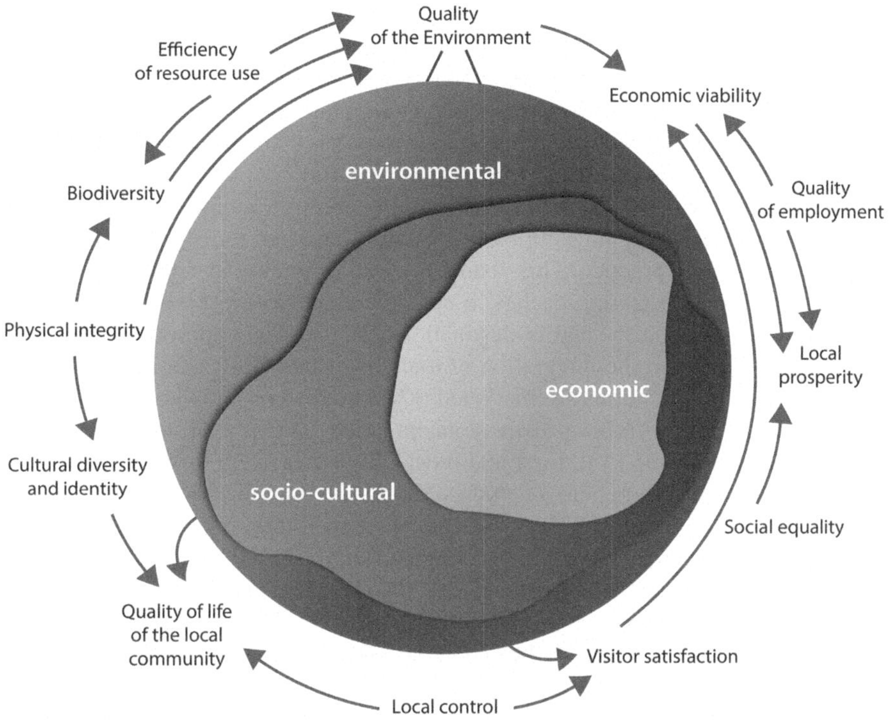

Fig. 2.2 Twelve key aspects of sustainable tourism. Source: Adapted from [57]:20; [42]

in other sectors, emphasising the positive role of social responsibility, is described by Ağan et al. [62]. This approach is close to the social responsibility applied by individual tourism actors with an emphasis on synergy, activation of social capital ([63]) and the spread of positive memes ([64]) within the network created. It can be called "collective social responsibility". Additionally, tourists can support through their selective demand the low-resource consumption. It has to be noted that ICT can promote social responsibility in tourism ([65]), but also undermine it. Examples of such undermining include the spread of distorted mental maps and stereotypes in social media, the mass presented offer of multinational and supra-regional companies suppressing the offer of local services, and the preference for digitised information over physical communication with local residents.

The system approach, as stated by Pásková [66] and Zelenka, Pásková and Husáková [42], comprehensively analyses the complex interplay of many external and internal factors, processes and elements of a tourism destination in the context of tourism sustainability. According to the mentioned authors, a tourism destination must always be considered as a very complex system open in time and space, where the destination is influenced by a number of external factors that are uncontrollable by destination management, and whose mutual interplay changes over time. According to Pásková [66], individual elements of this system can be analysed in terms of the nature of their action, intensity, positivity or negativity, activity or passivity. She explains that the element is either the agent or recipient of action in relation to other elements of the system. The system approach serves as a basis for various methods and approaches to modelling (simulating) tourism phenomena and their effects on the natural, economic and socio-cultural environment ([33, 42, 67, 68, 69, 70, 71]).

Over time, the vision of tourism sustainability, strongly supported by changing visitor preferences and changes in their lifestyles, has been transferred, discussed, analysed and put into practice. This vision was also beginning to significantly influence activities, mindsets, approaches to customers and ways of working together within the tourism industry and in destinations. The strategic approach of tourism stakeholders ([38]) and the integration of tourism sustainability into the way tourism services and destinations are marketed ([72]) are increasingly being promoted. An overview of the utilisation of the concept since 2002 is described by Torres-Delgado and Palomeque [73]. First and most comprehensively in the form of best practice guides and comprehensive studies, this vision is applied in concepts, strategies and practical measures within visitor management (see, e.g. [46, 48, 74]) and other approaches such as zonation, carrying capacity, LAC—Limits of Acceptable Change, access management, participatory management and integrated management, to tourism development in protected areas ([46, 48, 75, 76]). The varying extent to which different approaches (sectoral and organisational concepts, strategies to optimise the environmental impacts of tourism, sub-methods, techniques and activities) have been applied to tourism services is shown by the ways in which social responsibility has been applied in Sect. 2.2.

2.2 Causes and Factors of Un/sustainability of Tourism Development

The key to finding territorially specific solutions is to understand the causes, factors and mechanisms that lead to the fact that the development of tourism in a given region is not sustainable in the long term. The very unsustainability of tourism, or its rate is usually identified using extensive quantitative research, whilst understanding its causes requires intensive qualitative research. Based on previous results from case studies conducted in heavily visited destinations and other relevant published research (e.g. [7, 13, 14, 20, 21, 38, 77, 78, 79, 80, 81, 82, 83, 84, 85, 86, 87, 88]) three main causes of non/sustainability of tourism development can be recognised.

1. Tourism externalities and inequality in access to information in the process of social exchange

At the heart of the problem are the unreported and/or unintended impacts of tourism development in a given destination. In the planning or initiation phase of tourism development, its inhabitants are often not sufficiently informed or even have no awareness of its possible adverse effects on their quality of life. Part of these impacts are externalities that enter into the social exchange between guests and hosts. Some of the impacts of tourism are directly or indirectly caused by the activities of visitors (by their consumption of primary destination resources), but they are borne by local residents, whether from an economic, social or health point of view. If the development of tourism in a given destination is to be sustainable, the exchange between visitors and local residents has to be mutually beneficial and transparent. At the moment when the benefit from tourism income is outweighed by the impacts on the quality of life of local residents caused by the sharing of local public resources, the socio-cultural dimension of carrying capacity is exceeded ([30, 81, 83]). This leads, amongst other things, to a change in the price and orientation of the offer of local goods and services, responding more to the needs and possibilities of visitors than residents, to increased damage to the environment and the gradual commercialisation of public spaces and local traditions.

This problem is further aggravated by the fact that the so-called primary destination resources (natural and cultural tangible and intangible heritage of the region) are usually not renewable. These primary destination resources are often damaged by the development of secondary destination resources (tourism infrastructure). Not infrequently, so-called dual worlds emerge, the world of local residents, with a declining environment and crumbling social fabric on the one hand, and a tourist world offering experiences and luxury services to visitors on the other. If the problem of externalities and non-transparent informational environment of social exchange is not solved, it is not possible to achieve the sustainability of tourism development in the given destination.

2. Insufficient endogenous development of tourism in the given region

When evaluating the effect of tourism on the viability of a region, a continuum is offered, one extreme of which represents tourism as an activity complementary to the entrenched but adaptable activities of local actors, whilst the other extreme is tourism representing a substitute for the traditional activities of actors in a given region. In the first case, local entrepreneurs and non-entrepreneurs, through their involvement in the tourism industry in the given destination, earn extra money for their existing business or livelihood, thereby strengthening not only the multiplication of incomes, investments and employment, but also economic, ecological, social and cultural diversity of the region. In the second case, local economic activities are gradually displaced by tourism activities, which leads to a decrease in economic and other diversity (in extreme cases to a tourist monoculture) and subsequently to increased imports of goods and services into the region and associated economic leakages as well as economic dependence on the other regions. This second extreme results in a reduced level of economic self-sufficiency, weakening of cultural identity and damage to the ecosystems of the given region. Such development of tourism in the region is unsustainable in the long term.

3. The tragedy of commons leading to the so-called tourist trap

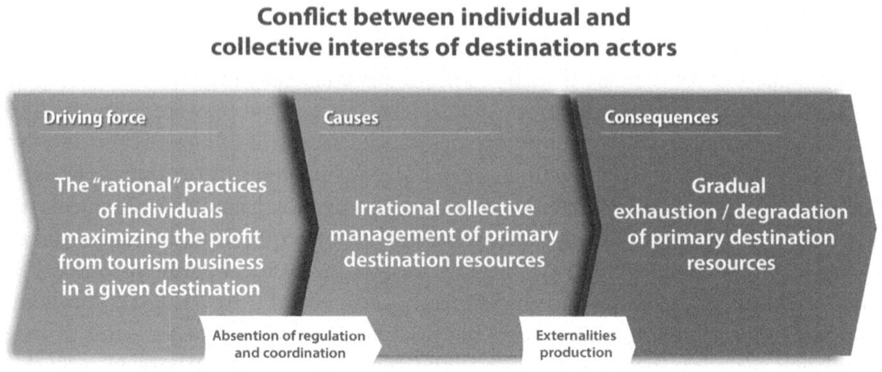

Fig. 2.3 Processual scheme of the tragedy of commons

The adverse effects of tourism development in the region are largely caused by the phenomenon known as the tragedy of commons ([89, 90]), when destination resources are used with such intensity and in such a way that they (often irreversibly) damage it. The problem is that the actors who share destination resources use them, compared to their private assets, not optimally, but maximally, disregarding their auto/regenerative capacity (Fig. 2.3).

The territories of tourism-exposed regions represent communal pasture and their efforts to sustainably manage resources depend on the quality of destination management, which is influenced by internal and external factors. The core of the problem of the tragedy of commons (hereinafter referred to as TOC) lies in the contradiction between individual and collective interests ([91]), whilst in the case of a tourism destination it is a contradiction between the interests of destination actors as individuals and the interests of the destination as a whole (destination actors as a collective). The resulting effect of the situation referred to as TOC is conditioned and influenced by the action of various factors. One of the conditioning factors is the fact that a significant part of the consumers of the primary destination resources are not residents of the destination, so the result of TOC action does not affect the resources on which depends their existence. These are mainly visitors and intermediaries of tourism services, such as travel agencies and tour operators. Furthermore, there is a multiplier effect of tourism, which is positively reflected in employment, income and investments in the region, but negatively in the depletion or damage of destination resources[3]. Jehlička, Tomeš and Daněk [91] cite the lack of a coordination mechanism as a fundamental conditioning factor, which Pásková [41] specifies as the absence of long-term functional destination management that would harmonise the individual interests of destination actors with their collective interests at the level of the entire destination. According to Jehlička, Tomeš and Daňek [91], the principle condition is the fact that all actors have unlimited access to resources. This

[3] Multiplication of emissions released into individual components of the environment as well as consumption of water, energy, food, fertilisers, building materials etc.

applies to a substantial part of destination resources, especially to the natural component of tourism potential.

2.3 Theoretical Concept of Sustainable Tourism

As emphasised in the literature ([38, 41, 48, 92, 93]), the basis of conceptual and sustainable tourism development is the use of approaches, techniques and results of basic and applied research in many scientific disciplines. Future research topics related to tourism sustainability are presented by Budeanu et al. [94]. The research into the application of tourism sustainability methods involves the creation of models and analyses. It also includes conceptual schemes, planning, monitoring and application of recommended strategic, tactical and reactive management activities to optimise tourism impacts. This research, like all research in the field of travel and tourism, is based on the interdisciplinary application of various disciplines, especially the humanities. The most important applications include:

- Sociology of tourism—Fundamental works by Cohen [1, 22, 95], Urry [96], MacCannell [97], the application of social exchange and the guest–host relationship by Doxey [23], Ap and Crompton [98] and Cohen and Cohen [99], the sociology of tourism as one of the theoretical bases of sustainable tourism by Pásková [38], and Wokoun and Vystoupil [100].
- Psychology of tourism—Key works by Cohen [22], Plog [101], Smith [102], Hahn and Kagelmann [103]; geopsychology with an emphasis on the experience of space and the relation to the landscape by Šípek [104] and Zelenka et al. [33]; psychology and sustainability of tourism by Venclová et al. [105], Han [106]; cognitive aspects of motivation and spacetime behaviour of visitors by Zelenka, Pásková and Husáková [48].
- Geography of tourism—Seminal works by Pearce [107], Ryan (1991), Williams and Shaw [108], Shaw and Williams [40]; spatial polarisation of tourism by Gilbert and Litt [109], Christaller [110], Shaw and Williams (1994); spatial dispersion of tourism by Miossec [111]; various aspects of geography of tourism and its sustainability by Pásková [38], Wokoun and Vystoupil [100], Mariot [112].
- Tourism economics—Principal works focusing, in particular, on the destination life cycle by Christaller [110], Noronha [25], Butler [9], Weaver [113, 114], Pásková [38, 82], Song et al. [49]; tourism economics including environmental economics by Garrod and Fyall [115].
- Anthropology of tourism—Basic work by Smith [116], studies on Indigenous knowledge by Butler and Hinch [117], Pásková [15, 38, 118]; Konopíková and Soukup [119], Carr, Ruhanen and Whitford [120], Wu, Wall and Tsou [121], Půtová [122].
- Philosophy of tourism—Substantial work by Tribe [123]; approaches to research and the concept of the tourism phenomenon by Tribe and Xiao [124].

- Tourism marketing—Primary works by Morrison [125], Middleton [126], Kotler, Bowen and Makens [127]; marketing development directions by Hu [128], Palatková [36], Zelenka [129].
- Computer science and cognitive science in tourism—Core works by Buhalis and Law [130]; ICT and tourism sustainability including learning tourism destination, Internet of Things, AI and system dynamics modelling by Schianetz, Kavanagh and Lockington [131], Ali and Frew [60], Zelenka et al. [74], Zelenka, Pásková and Husáková [48], Lei, Wei, Otilia [132].
- Environmental tourism—Major works by Turner and Ash [133], Rosenow and Pulsipher [37], Heukemes [134], Weaver [135], Pásková [38, 41] and Zelenka et al. [93].

In order for tourism to be sustainable, theoretical concepts need to be further developed with an emphasis on a deeper understanding of the regularities and patterns of the phenomena they describe and possibly predict. Part of developing these nomothetic approaches is to critically analyse them, as shown in the literature, for example for the concept of destination life cycle ([13, 14, 38, 136, 137]), for the concept of carrying capacity ([39, 76, 83, 137, 138, 139, 140]) and for the very feasibility of implementing sustainable tourism development ([60]: 45–46, [135, 141, 142, 143]). In addition to the theoretical concepts of sustainable tourism (for an up-to-date overview of how they are developed see [144]), their practical applications are also developed. For example, the ISO organisation declares a systematic support for tourism ([145]), for example the ISO 20121 ([146]) for the sustainable system management of major events. Formalised evidence to support the practical application of sustainable tourism practices thus includes methodologies, best practice guides, systemic approaches, standards and elaborated case studies. They primarily focus on:

- Tourism carrying capacity of a particular area, the capacity of the infrastructure, the tolerance of the local population and the limits of the organisational infrastructure for tourism development:
  - The concept of carrying capacity ([1, 23, 38, 39, 41, 82, 137, 147]).
  - The link between tourism carrying capacity and the life cycle of a destination ([9, 23, 38, 41, 82, 148]).
  - Destination- and site-specific carrying capacity—e.g. for caves Lobo et al. [140], modelling for national parks Vujko et al. [47].
- Monitoring the impacts of tourism:
  - Monitoring techniques and management monitoring ([93, 149, 150, 151, 152, 153, 154]).
  - Modelling of tourism impacts—Erosion, emissions, congestion etc., e.g. Saenz-de-Miera and Rosselló [155] simulated emissions associated with tourism in Mallorca.
- Perception of tourism impacts:

- Public attitudes towards tourism impacts—Understanding the type of tourism impacts and their severity is important, but as shown by, for example Miller et al. [156] for England, the public is not yet well informed about these impacts and appears reluctant to change their tourism participation behaviour.
- The severity of environmental problems perceived by tourism participants—An overview of perceptions of different types of tourism impacts was presented by Stefănica and Butnaru [157].
- Perception of tourism impact on the authenticity of a destination—A frequently studied issue from the point of view of local residents and visitors to the destination ([17, 38, 66, 102, 147, 158, 159, 160, 161, 162]), maintaining this authenticity represents one of the pillars of the tourism sustainability.
- A review study of methods of research on the perception of tourism impacts on residents' quality of life was conducted by Sharpley [163], Wang and Chen [164] emphasised the importance of residents' perception of local identity in their perception of tourism impacts, and Woo, Kim, and Uysal [165] analysed the effects of tangible and intangible aspects of tourism development on residents' perception of quality of life and their attitude towards tourism promotion.

• Systemic approach to the carrying capacity effects of tourism on a tourism destination is emphasised as a necessary theoretical starting point ([17, 38, 66, 102, 147, 158, 159, 160, 161, 162]) and as the basis for practical applications of carrying capacity and the concept of destination life cycle ([9, 25, 66, 83, 93, 113, 166]).
• Approaches for tourism sustainability management:

  - Destination management includes tools, approaches and concepts such as visitor management, participatory management, social capital, management of social, economic and environmental impacts of tourism ([38, 93, 167, 168]).
  - Visitor management ([169, 170, 171, 172, 173]).
  - Management of social impacts of tourism ([38, 82, 174]).
  - Participatory tourism management—fundamental works by Eagles, Mccool and Haynes [150], COMMITTEE [175], applications in protected areas with respect to the CBD (the Convention on Biological Diversity), which operates under the United Nations Environment Programme (UNEP) ([176]), for cooperation of key actors see Beritelli [177], for summary and discussion of various approaches see Waligo, Clarke and Hawkins [68], Zelenka et al. [93], for discussion of issues of participation of tourism actors see Hatipoglu, Alvarez and Ertuna [178].
  - Participatory tourism planning—As stated, for example by Haywood [179] and COMMITTEE [175], the intention of participatory tourism management is the progressively increasing participation of tourism actors, including the local community, in tourism planning and development; residents' perceptions of their own quality of life as influenced by tourism [180] and a comparison of the positive and negative impacts of tourism as perceived by residents [17] are also important for promoting participatory management.

  - Participatory tourism decision-making—Similarly, participatory tourism management involves the inclusion of the local community and Indigenous peoples in tourism development decision-making, which has been critically analysed by Joppe [181] and the link between decision-making and community-based tourism development scenarios has been presented by Okazaki [182]; see also community-based tourism in Sect. 6.3.
  - Participatory tourism research and development—The inclusion of traditional knowledge of local and Indigenous communities in tourism research and development and destination management is an important component of various forms of sustainable tourism—geotourism, community-based tourism, as described by, e.g. Pásková and Dowling [183], Pásková [118, 184]).

- Reduction in consumption of non-renewable and renewable resources, recycling and use of renewable energy sources:
  - Reducing fossil fuel consumption and CO2 emissions (e.g. [185, 186, 187, 188]).
  - Eco-certification—e.g. Gössling and Buckley [187].

- Holistic and systemic approaches to tourism management ([38, 66, 93, 189]), including appropriate flexibility in the concept of sustainable tourism based on local conditions ([83, 189]).

Another very important issue is the discussion of the possibilities of targeted development of sustainable tourism and the destination-specific verification of the related theoretical concepts, focusing on capturing the destination development and the possibilities of its prediction and control should be done through the application of the concept of destination life cycle and other models of destination development, as described, for example by Weaver [114] and Pásková [38, 82]. The conditions for the potential transformation of unsustainable mass tourism into sustainable mass tourism were proposed conceptually by Weaver [114, 142, 143] and assessed critically by Peeters [141]. Farrell and Twining-Ward [190] as well as many other authors (e.g. [38, 41, 191]), with respect to the interdisciplinary nature of the issue of conceptual and sustainable tourism development, have emphasised that an interdisciplinary and holistic approach to research is necessary for the transition to sustainable tourism. It is therefore desirable to use expertise in, amongst other areas, ecosystem ecology, environmental economics, global science, sustainability science with an emphasis on global change and systemic approaches with a focus on the complexity of the problem. Such a broadly inclusive approach to sustainable tourism also helps to define the role and specific contributions of tourism in achieving the 17 UN Sustainable Development Goals (e.g. [192, 193, 194]).

The following clichés and professional inaccuracies are gradually overcome:

- Incompatibility of tourism development with nature and landscape protection:
  - For clarification of the invalidity of such clichés see the concept and methodology of ecotourism and geotourism development ([93, 195]).

  - Gradually changed concepts of Protected Areas (PAs) in relation to tourism ([45, 93, 196, 197, 198]).
  - The mutual dependency between biodiversity and tourism ([59, 75, 199]).
  - Optimisation and participatory approaches to tourism management in protected areas ([83, 93, 150]).

- Conflict between economic and conservation interests

  - Clarifying the irrelevance of these clichés ([38, 41, 150, 200]).
  - Discussion on the importance of tourism for financing the operation of protected areas and preservation of its fauna, flora and fragile ecosystems ([93, 150]).
  - Recommended approaches to funding protected areas ([93, 150, 200, 201]).
  - Discussion on the relationship between tourism and protected areas ([93, 150, 200]).
  - Comprehensive system analysis of tourism revenue, sustainability and tourism development planning in Northcote and Macbeth [202].

- Inevitable conflicts between the interests of visitors and locals—Clarification of the invalidity of this statement is based on the application of the concept of sustainable forms of tourism (ecotourism, community-based tourism, Indigenous tourism, geotourism etc.; [195]) in Chap. 6.

Increasingly and with accelerating success, literature has focused on the search for harmony, synergy between sustainable tourism and nature and landscape conservation ([38, 59, 93, 150, 203]) and between tourism development and the interests of local people ([83, 117, 118, 204]). Holistic approaches have been developed to optimise tourism performance in different types of areas, e.g. for protected areas by Eagles, Mccool and Haynes [150], Phillips [45], EUROPARC Federation [196] and Zelenka et al. [93] and for geoparks by Pásková [38]).

To plan and implement activities improving tourism sustainability, it is essential to search for destination-specific practices, defined by systematic research. Thus, Tosun [205], for example examined the specific features of sustainable tourism in developing countries.

2.4 Determination of Tourism Sustainability and Social Responsibility

During the twentieth century, tourism underwent a period of extensive growth and at the same time, in addition to fundamental technological changes (ICT, technology in transportation and accommodation), changes in the interpretation of natural and cultural heritage and in marketing, changes in basic paradigms, including the meaning of travel and related ethics. A transition from anthropogenic ethics (anthropocentric approach) to environmental ethics (biocentric approach) with an emphasis on respecting the environment is necessary, e.g. [206, 207]. There were also fundamental changes in the perception of negative impacts and/or benefits to the local community, economy and environment. As a rule, the local community bears all the negative consequences of tourism development, but receives only part of its benefits (e.g. Joppe 1996). All these changes have led to a reassessment of the meaning, forms and degree of cooperation amongst tourism actors (e.g. the model of cooperation in tourism in [208]; the analysis of conditions and forms of participation in [175]) and the gradual formation of destination management theory and practice. This development was marked for a relatively long time by a low level of social responsibility of investors in tourism infrastructure, tourism service providers, destination management organisations, tourism service intermediaries, visitors, national and regional politicians and other tourism actors. The change in the approach to fulfilling social responsibility in tourism occurred gradually since the 1970s and included sustainability in terms of its management, scientific approaches, time dimension, monitoring, planning and the development of many other approaches and tools. In recent years, sustainability of tourism has been increasingly perceived in marketing as the basis for successful communication with (potential) tourism participants as well as other tourism actors, local authorities and state nature and landscape conservation authorities. The greatest potential is available for destinations and tourist attractions if their administration (destination management organisations) recognise the importance of the long-term comparative and competitive advantages of a sustainable tourism development strategy and if sustainability of tourism is properly linked (not only) in marketing terms with the current trend of growing awareness of social responsibility in the value system of individuals and society as a whole.

Fundamental to the promotion of approaches to sustainability of tourism has been its definition, which is difficult. The definition of sustainable tourism is even at the global level (UNWTO) gradually being modified and refined to better apply the general concept of sustainable development to tourism. The current UNWTO definition of sustainable tourism is based on a consensus and responsibility, assumed and progressively fulfilled within the framework of sustainable tourism ([209]):

> Tourism that takes full account of its current and future economic, social and environmental impacts, addressing the needs of visitors, the industry, the environment and host communities.

Such responsibility can also be expressed ([92, 210]) in relation to all components of the environment in terms of process and planned objectives: tourism, which, thanks to conceptual planning and development, does not disturb the natural, cultural and social environments in the long term, as its aim is to preserve the quality of the environment in all its aspects and respect the lifestyle (including the existing livelihood) of the local population.

In the pursuit of sustainable tourism, the basic territorial unit is the tourism destination. Sustainable tourism, according to Pásková [38] and Zelenka, Pásková and Husáková [48], in terms of destination management, as seen in Fig. 2.4, is mainly achieved through appropriate forms of management, marketing, optimisation

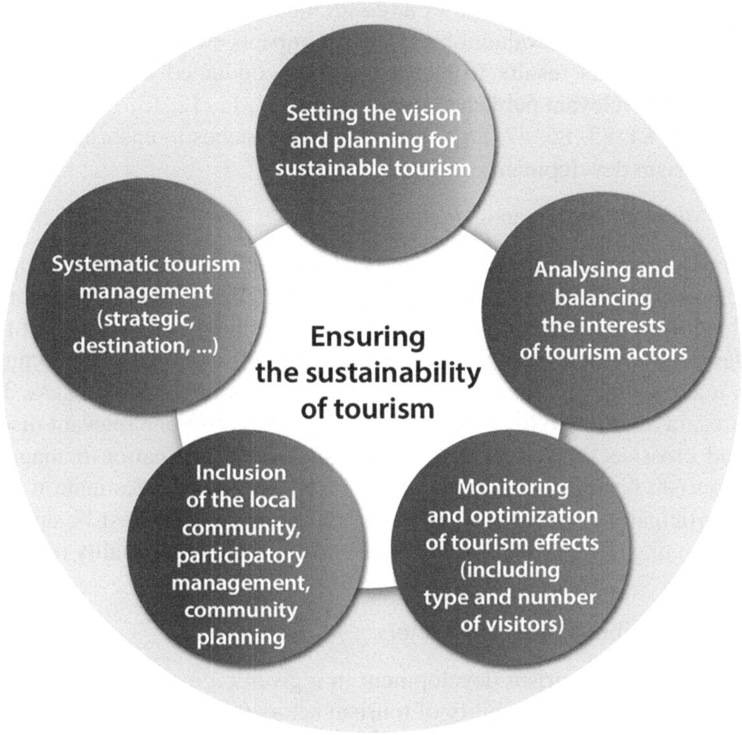

Fig. 2.4 The most important features and aspects of implementing sustainable tourism. Source: Inspired by [74]:11

techniques and the development of appropriate forms of tourism in accordance with the preservation and development of the destination's resources and their authenticity. According to Mihalič [211], the use of environmental management in destination management increases the competitiveness of a destination.

2.5 Approaches to Achieve Sustainability of Tourism Development

The basic approaches to achieving the sustainability of tourism development ([41, 83]) include various types of instruments of a normative nature (legislation, regulations, methodologies etc.), of an economic nature (price regulation, incentives etc.), of a conceptual and managerial nature (strategic management, visitor management, access management etc.), of an informative, educational and educational nature, of a voluntary nature (voluntary commitments and activities, certification schemes), of an institutional nature (destination management organisations, tourist microregions, professional associations etc.) and of a research nature (activity of external panels, monitoring, analysis and evaluation of tourism impacts etc.).

Based on previous results from case studies conducted in intensive tourism regions and other relevant published research (e.g. [7, 11, 12, 13, 14, 20, 21, 23, 41, 80, 81, 82, 83, 84, 85, 86, 87, 88]) three main approaches to ensuring the sustainability of tourism development can be pointed out.

1. Quality of destination management

The destination management team should be based on interdisciplinary cooperation and should use the results of tourism research, especially the continuous monitoring and evaluation of its impact on the given region. Optimising the effects of tourism on the development of a given region, which leads to the best possible compromise between maximising benefits and minimising risks, is a long-term process. The latter requires, in addition to deep expertise and experience in one relevant discipline, also broad cross-sectional knowledge and skills from destination managers or a specific tourism business (T-type manager, [168]). As part of destination management, a participatory approach to research and management must be applied and strategic planning, knowledge management and environmental quality management must be used.

2. Social responsibility of tourism actors

The sustainability of tourism development in a given region is largely determined by the level of social responsibility of tourism actors (e.g. [212]). It is essential that these actors, both on the demand and supply side, become aware of their own contribution to the adverse effects of tourism on the environment and the quality of life of the inhabitants of the given region. Subsequently, it is necessary that these actors accept social responsibility for their contribution to the given impact of tourism on the community and the destination environment. On the demand side, it concerns the responsible consumer behaviour of visitors, specifically their choice to prioritise the products of responsible operators of tourism services. This signals to the tourism market a change in the requirements for its new dimension of quality—sustainability. This signal becomes significant if this type of consumer forms a recognizable segment of the tourism industry. A change in needs, values and purchasing preferences on the demand side of tourism can thus lead to a gradual transformation of the supply. It is a positive fact that entrepreneurs in the tourism industry are gradually beginning to understand the responsibility for the undesirable effects of their business on the quality of life of the given region as an opportunity to differentiate themselves from the competition with an offer intended for the growing segment of responsible participants in the tourism industry ([213, 214]).

3. System concept of environmental management performance

Due to the increasing number of problems arising from the irrational treatment of natural resources, the aspect of sustainability no longer represents an externally imposed limitation for businesses and other entities, but rather the possibility of their future development and application of innovations ([44]). Reducing the ecological footprint appears to be one of the approaches to optimising the effects of tourism on the destination environment. If tourism actors are adequately informed, this responsible behaviour can lead not only to an increase in the competitiveness of the entity or the entire destination but also to a reduction in the costs of tourism development. It is mainly about the possibilities and tools of environmental management, such as the use of environmentally friendly energies and technologies as well as the application of circular economy principles in the development of activities and the construction of tourism infrastructure in a given region[4]. Preventive approaches, continuous work on improving processes in the organisation, building environmental quality networks and a multi-level system of audits are important. A specific approach is offered by the concept of comprehensive management of environmental quality (e.g. [215]), whose fundamental principles include a systemic approach, the participation of all employees of the organisation, the prevention of environmental risks and damages, and the continuous growth of the quality of the environmental profile the organisation, including the management of its processes.

References

1. Cohen, E. (1979). Rethinking the sociology of tourism. *Annals of Tourism Research, 6*(1), 18–35. https://doi.org/10.1016/0160-7383(79)90092-6
2. WTAY. (2014). WTAY code of conduct, [online]. Retrieved August 22, 2022, from http://wtay.com/resources/code-of-conduct/
3. Pásková, M., & Zelenka, J. (2016). Social responsibility role in tourism sustainability. In P. Jedlička (Ed.), *Proceedings of Hradec economic days, February 2nd and 3rd*, ISSN 2464-6059 (pp. 777–785).
4. Tosun, C. (1998). Roots of unsustainable tourism development at the local level: The case of Urgup in Turkey. *Tourism Management, 19*(6), 595–610. https://doi.org/10.1016/S0261-5177(98)00068-5
5. Coria, J., & Calfucura, E. (2012). Ecotourism and the development of indigenous communities: The good, the bad, and the ugly. *Ecological Economics, 73*, 47–55. https://doi.org/10.1016/j.ecolecon.2011.10.024
6. Raybould, M. Anning, D., Ware, D., & Lazarow, N. (2013). Beach and surf tourism and recreation in Australia: Vulnerability and adaptation. Fisheries research and development corporation a bond university, p. 241, ISBN: 978-0-646-90467-2 [online]. Retrieved January 14, 2024, from https://www.frdc.com.au/sites/default/files/products/2010-536-DLD.pdf
7. Agarwal, S. (1994). The resort cycle revisited: Implications for resorts. In C. P. Cooper & A. Lockwood (Eds.), *Progress in tourism, recreation and hospitality management* (Vol. 5, pp. 194–208). Wiley.
8. Butler, R. V. (2006). *Aspects of tourism. The tourism area life cycle: Applications and modifications* (p. 385). Cromwell Press Limited.
9. Butler, R. W. (1980). The concept of a tourism area cycle of evolution: Implication for management of resources. *Canadian Geographer, 24*(1), 5–12. https://doi.org/10.1111/j.1541-0064.1980.tb00970.x

[4]Use of renewable energy sources, prevention and sorting of waste, use of low-energy appliances and means of transport, use of operation outputs as its inputs etc.

10. Cole, S. (2012). Synergy and congestion in the tourist destination life cycle. *Tourism Management, 33*(5), 1128–1140. https://doi.org/10.1016/j.tourman.2011.11.018
11. Cooper, C. (1994). The destination life cycle: An update. In A. V. Seaton (Ed.), *Tourism: The state of the art* (pp. 340–346). John Wiley & Sons.
12. Getz, D. (1992). Tourism planning and destination life cycle. *Annals of Tourism Research, 19*(4), 752–770. https://doi.org/10.1016/0160-7383(92)90065-W
13. Haywood, M. K. (1986). Can the tourist-area life cycle be made operational? *Tourism Management, 7*(2), 154–167. https://doi.org/10.1016/0261-5177(86)90002-6
14. Hovinen, G. R. (1981). A tourist cycle in Lancaster Country, Pennsylvania. *Canadian Geographer, 15*(3), 283–286. https://doi.org/10.1111/j.1541-0064.1981.tb01649.x
15. Pásková, M. (2022). Geopark certification as an efficient form of sustainable management of a geotourism destination. In V. Braga, A. Duarte, & C. S. Marques (Eds.), *Economics and management of geotourism. Tourism, hospitality & event management* (pp. 65–85, p. 124). Springer. https://doi.org/10.1007/978-3-030-89839-7_4
16. Priestley, G., & Mundet, L. (1998). The post-stagnation phase of the resort cycle. *Annals of Tourism Research, 25*(1), 85–111. https://doi.org/10.1016/S0160-7383(97)00062-5
17. Shaw, G., & Williams, A. M. (1994). *Critical issues in tourism: A geographical perspective* (p. 280). Blackwell Publishers.
18. Wall, G. (1982). Cycles and capacity: Incipient theory or conceptual contradiction? *Tourism Management, 3*(3), 188–192. https://doi.org/10.1177/004728758302200158
19. Wall, G. (1983). Cycles and capacity: A contradiction in terms. *Annals of Tourism Research, 10*(2), 268–270. https://doi.org/10.1016/0160-7383(83)90030-0
20. Brougham, J. E., & Butler, R. W. (1981). A segmentation analysis of resident attitudes to the social impact of tourism. *Annals of Tourism Research, 8*(4), 569–589. https://doi.org/10.1016/0160-7383(81)90042-6
21. Brunt, P., & Courtney, P. (1999). Host perceptions of sociocultural impacts. *Annals of Tourism Research, 26*(3), 493–515. https://doi.org/10.1016/S0160-7383(99)00003-1
22. Cohen, E. (1972). Towards a sociology of international tourism. *Social Research, 39*(1), 164–182. https://www.jstor.org/stable/40970087
23. Doxey, G. V. A. (1975). Causation theory of visitor-resident irritants; methodology and research inference. The Travel Research Association Conference No. 6. San Diego: TTRA, pp. 195–198.
24. Kim, K., Uysal, M., & Sirgy, M. J. (2013). How does tourism in a community impact the quality of life of community residents? *Tourism Management, 36*, 527–540. https://doi.org/10.1016/j.tourman.2012.09.005
25. Murphy, P. E. (1985). *Tourism: A community approach* (p. 218). Methuen.
26. Pásková, M., Wall, G., Zejda, D., & Zelenka, J. (2021). Tourism carrying capacity reconceptualization: Modelling and management of destinations. *Journal of Destination Marketing & Management, 21*, 100638. https://doi.org/10.1016/j.jdmm.2021.100638
27. Plog, S. C. (1977). Why destination rise and fall in popularity. In E. M. Kelly (Ed.), *Domestic and international tourism* (pp. 26–28). Institute of Certified Travel Agents.
28. Plog, S. C. (1974). Why destinations rise and fall in popularity. *Cornell Hotel and Restaurant Administration Quarterly, 14*(4), 13–16.
29. Russo, A. P. (2002b). A stakeholders' approach to tourism policy in Bruges. In M. Jansen–Verbeke (Ed.), *Tourism studies in Bruges* (pp. 32–41). WES Research and Consultancy.
30. Butler, R. W. (2019). Tourism carrying capacity research: A perspective article. *Tourism Review, 75*(1), 207–211. https://doi.org/10.1108/TR-05-2019-0194
31. McCool, S. F., & Lime, D. W. (2001). Tourism carrying capacity: Tempting fantasy or useful reality. *Journal of Sustainable Tourism, 9*(5), 372–388. https://doi.org/10.1080/09669580108667409
32. Wall, G. (2020). From carrying capacity to overtourism: A perspective article. *Tourism Review, 75*(1), 212–215. https://doi.org/10.1108/TR-08-2019-0356
33. Young, G. (1973). *Tourism (blessing or blight?)* (p. 191). Penguin Books.

34. Buckley, R. (2012). Sustainable tourism: Research and reality. *Annals of Tourism Research, 39*(2), 528–546. https://doi.org/10.1016/j.annals.2012.02.003
35. Ballantyne, R., Packer, J., & Hughes, K. (2009). Tourists' support for conservation messages and sustainable management practices in wildlife tourism experiences. *Tourism Management, 30*(5), 658–664. https://doi.org/10.1016/j.tourman.2008.11.003
36. Juvan, E., & Dolnicar, S. (2014). The attitude–behaviour gap in sustainable tourism. *Annals of Tourism Research, 48*, 76–95. https://doi.org/10.1016/j.annals.2014.05.012
37. Reisinger, Y., & Steiner, C. (2006b). Reconceptualising interpretation: The role of tour guides in authentic tourism. *Current Issues in Tourism, 9*(6), 481–498. https://doi.org/10.2167/cit280.0
38. Park, S.-Y., & Lee, S. (2009). Financial rewards for social responsibility. A mixed picture for restaurant companies. *Cornell Hospitality Quarterly, 50*(2), 168–179. https://doi.org/10.1177/1938965509331814
39. Zahra, A., & McGehee, N. G. (2013). Volunteer tourism: A host community capital perspective. *Annals of Tourism Research, 42*, 22–45. https://doi.org/10.1016/j.annals.2013.01.008
40. Weaver, D. B. (2000). A broad context model of destination development scenarios. *Tourism Management, 21*(3), 217–224. https://doi.org/10.1016/S0261-5177(99)00054-0
41. Palmer, T., & Riera, A. (2003). Tourism and environmental taxes. With special reference to the "Balearic ecotax". *Tourism Management, 24*(6), 665–674. https://doi.org/10.1016/S0261-5177(03)00046-3
42. Zein, K., et al. (2008). Best environmental practices for the hotel industry. Sustainable Business Associates (sba), p. 56, [online]. Retrieved January 14, 2024, from https://nrcne.org/wp-content/uploads/2019/12/SBABGEHOTELLERIEENG2008.pdf
43. ISO 21416:2019. Recreational diving services. Requirements and guidance on environmentally sustainable practices in recreational diving.
44. Hart, S. L. (1992). An integrative framework for strategy-making processes. *The Academy of Management Review, 17*(2), 327–351.
45. Pásková, M., & Hradecký, P. (2014). Aspiring Geopark Rio Coco (Nicaragua). In Conference Proceedings of 14th International Multidisciplinary Scientific Geoconference SGEM 2014, Geoconference on Ecology, Economics, Education and Legislation. Ecology and Environmental Protection. Albena, Bulgaria, Vol. II, pp. 53–60, ISBN 978-619-7105-18-6.
46. easyJet. (2015). Presentation of Corporate Social Responsibility on the easyJet, plc website in 2015, [online]. Retrieved August 21, 2015, from http://corporate.easyjet.com/corporate-responsibility.aspx?sc_lang=en
47. Urry, J. (1990). *The tourist gaze: Leisure and travel in contemporary society (theory, culture and society series)* (p. 184). SAGE Publications. ISBN 978-0803981836.
48. Zelenka, J. (2001). Integrated management. *E+M Ekonomie a Management, 4*(1), 11–14.
49. Singh, S. (2011). The tourism area 'life cycle': A clarification. *Annals of Tourism Research, 38*(3), 1185–1187. https://doi.org/10.1016/j.annals.2011.03.005
50. Font, X., & Lynes, J. (2018). Corporate social responsibility in tourism and hospitality. *Journal of Sustainable Tourism, 26*(7), 1027–1042. https://doi.org/10.1080/09669582.2018.1488856
51. Wang, Y.-F., Chen, S.-P., Lee, Y.-C., & Tsai, C.-T. (2013). Developing green management standards for restaurants: An application of green supply chain management. *International Journal of Hospitality Management, 34*, 263–273. https://doi.org/10.1016/j.ijhm.2013.04.001
52. Rastegar, R., & Ruhanen, L. (2023). Climate change and tourism transition: From cosmopolitan to local justice. *Annals of Tourism Research, 100*, 103565. https://doi.org/10.1016/j.annals.2023.103565
53. Cernat, L., & Gourdon, J. (2007). *Is the concept of sustainable tourism sustainable? Developing the sustainable tourism benchmarking tool*. ISSN 1816 – 2878 (p. 27).
54. Council of National Geoparks. (2017). Charta národních geoparků České republiky [Charter of National Geoparks of the Czech Republic], [online]. Retrieved December 22, 2017, from http://www.geology.cz/narodnigeoparky/rada-ngcr

55. Tourism Sustainability Group. (2007). Action for more Sustainable European Tourism. Report of the Tourism Sustainability Group. http://www.ernestproject.eu/coalap/ernest/other_docs/10072009160632789_tsg_final_report_en.pdf
56. Hayes-Conroy, A. (2010). Feeling slow food: Visceral fieldwork and empathetic research relations in the alternative food movement. *Geoforum, 41*(5), 734–742. https://doi.org/10.1016/j.geoforum.2010.04.005
57. Tourisme Québec. (2017). The ecotourist code of conduct, [online]. Retrieved June 22, 2017, from https://www.tourisme.gouv.qc.ca/publications/media/document/etudes-statistiques/code_eco_a.pdf
58. WWF. (2022). Code of conduct for Arctic Tourists, WWF International Arctic Programme, [online]. Retrieved August 10, 2022, from https://wwfeu.awsassets.panda.org/downloads/codeofconductforarctictourists(eng).pdf
59. Turner, L., & Ash, J. (1975). *The golden hordes* (p. 319). Constable.
60. Ali, A., & Frew, A. J. (2012). *Information and communication technologies for sustainable tourism* (1st ed.). Routledge. ISBN 978-0415673174.
61. Lovelock, B. (2008). Ethical travel decisions travel agents and human rights. *Annals of Tourism Research, 35*(2), 338–358. https://doi.org/10.1016/j.annals.2007.08.004
62. Ağan, Y., et al. (2016). The relationships between corporate social responsibility, environmental supplier development, and firm performance. *Journal of Cleaner Production, 112*(3), 1872–1881. https://doi.org/10.1016/j.jclepro.2014.08.090
63. Zelenka, J., et al. (2008). *Percepce krajiny a genius loci [Perception of the landscape and Genius Loci]* (p. 326). Gaudeamus Hradec Králové. ISBN 978-80-7041-191-9.
64. Blackmore, S. (2000). *The meme machine* (p. 264). Oxford University Press. ISBN 0-19-286212-X.
65. Mostafanezhad, M. (2014). Volunteer tourism and the popular humanitarian gaze. *Geoforum, 54*, 111–118. https://doi.org/10.1016/j.geoforum.2014.04.004
66. Palatková, M. (2012). Travel and tourism marketing – Review. *Czech Journal of Tourism, 1*(1), 30–52.
67. Marzouki, M., Froger, G., & Ballet, J. (2012). Ecotourism versus mass tourism. A comparison of environmental impacts based on ecological footprint analysis. *Sustainability, 4*(1), 123–140. https://doi.org/10.3390/su4010123
68. Václav Havel Airport Prague. (2023). Prague airport's development quest: Sustainable and Gree, [online]. Retrieved May 26, 2023, from https://www.prg.aero/en/letiste-praha-se-rozviji-udrzitelne-do-zelenych-barev
69. Varley, P., & Medway, D. (2011). Ecosophy and tourism: Rethinking a mountain resort. *Tourism Management, 32*(4), 902–911. https://doi.org/10.1016/j.tourman.2010.08.005
70. Wokoun, R., & Vystoupil, J. (1987). *Geografie cestovního ruchu a rekreace I [Geography of tourism and recreation I]* (p. 250). Státní pedagogické nakladatelství.
71. Woo, E., Kim, H., & Uysal, M. (2015). Life satisfaction and support for tourism development. *Annals of Tourism Research, 50*, 84–97. https://doi.org/10.1016/j.annals.2014.11.001
72. Yilmazdogan, O. C., Secilmis, C., & Cicek, D. (2015). The effect of Corporate Social Responsibility (CSR) perception on tourism students' intention to work in sector. *Procedia Economics and Finance, 23*, 1340–1346. https://doi.org/10.1016/S2212-5671(15)00321-4
73. Thomlinson, E., & Getz, D. (1996). The question of scale in ecotourism: Case study of two small ecotour operators in the Mundo Maya region of Central America. *Journal of Sustainable Tourism, 4*(4), 183–188. https://doi.org/10.1080/09669589608667267
74. Young, M., Higham, J. E. S., & Reis, A. C. (2014). 'Up in the air': A conceptual critique of flying addiction. *Annals of Tourism Research, 49*, 51–64. https://doi.org/10.1016/j.annals.2014.08.003
75. CBD. (2004). Guidelines on biodiversity and tourism development. Montreal: Secretariat of the Convention on Biological Diversity, p. 34, ISBN 92-807-2468-1, [online]. Retrieved August 15, 2022, from https://www.cbd.int/doc/publications/tou-gdl-en.pdf

76. Hammitt, W. E., & Cole, D. N. (1998). *Wildland Recreation: Ecology and Management* (2nd ed., p. 361). John Wiley.
77. Coppock, J. T. (1977). *Second homes: Curse or blessing* (p. 241). Pergamon Press.
78. Dredge, D. (2022). Regenerative tourism: Transforming mindsets, systems and practices. *Journal of Tourism Futures, 8*(3), 269–281. https://doi.org/10.1108/JTF-01-2022-0015
79. GGN. (2022). Code of ethics, global geoparks network, [online]. Retrieved August 4, 2022, from https://globalgeoparksnetwork.org/wp-content/uploads/2016/07/GLOBAL-GEOPARKS-NETWORK-CODE-OF-ETHICS-final.pdf
80. Kang, K. H., Lee, S., & Huh, C. (2010). Impacts of positive and negative corporate social responsibility activities on company performance in the hospitality industry. *International Journal of Hospitality Management, 29*(1), 72–82. https://doi.org/10.1016/j.ijhm.2009.05.006
81. Martínez, P., & del Bosque, I. R. (2013). CSR and customer loyalty: The roles of trust, customer identification with the company and satisfaction. *International Journal of Hospitality Management, 35*, 89–99. https://doi.org/10.1016/j.ijhm.2013.05.009
82. Palhares, J. C. P., & Pezzopane, J. R. M. (2015). Water footprint accounting and scarcity indicators of conventional and organic dairy production systems. *Journal of Cleaner Production, 93*, 299–307. https://doi.org/10.1016/j.jclepro.2015.01.035
83. Pásková, M. (2014). *Udržitelnost cestovního ruchu [Tourism sustainability]* (p. 335). Gaudeamus Hradec Králové. ISBN 978-80-7435-329-1.
84. Pásková, M., Zelenka, J., Ogasawara, T., Zavala, B., & Astete, I. (2021b). The ABC concept – Value added to the earth heritage Interpretation? *Geoheritage, 13*, 38. https://doi.org/10.1007/s12371-021-00558-8
85. Peeters, P. (2012). A clear path towards sustainable mass tourism? Rejoinder to the paper 'Organic, incremental and induced paths to sustainable mass tourism convergence' by David B. Weaver. *Tourism Management, 33*(5), 1038–1041. https://doi.org/10.1016/j.tourman.2012.01.013
86. Reisinger, Y., & Steiner, C. J. (2006a). Reconceptualizing object authenticity. *Annals of Tourism Research, 33*(1), 65–86. https://doi.org/10.1016/j.annals.2005.04.003
87. Report. (1998). International conference on sustainable tourism in Small Island Developing States (SIDS) and other Islands, Lanzarote, Spain, 25–28 October 1998, final report, p. 57.
88. Vourc'h, A., & Denman, R. (2003). Tourism and Local Agenda 21. The role of local authorities in sustainable tourism. UN, [online]. Retrieved June 14, 2013, from http://www.unep.fr/shared/publications/pdf/-3207-Tourism-Agenda.pdf, ISBN: 92-807-2267-0.
89. Hardin, G. (1968). The tragedy of the commons. *Science, 162*(3859), 1243–1248. https://doi.org/10.1126/science.162.3859.1243
90. Hardin, G. (1991). The tragedy of the unmanaged commons. Population and the disguises of providence. In R. V. Andelson (Ed.), *Commons without tragedy: Protecting the environment from over-population. A new approach* (pp. 162–185). Shepheard-Walwyn Publishers.
91. ISO 50001:2018 Energy management systems – Requirements with guidance for use.
92. Zambrano, A. M. A., Broadbent, E. N., & Durham, W. H. (2010). Social and environmental effects of ecotourism in the Osa Peninsula of Costa Rica: the Lapa Rios case. *Journal of Ecotourism, 9*(1), 62–83. https://doi.org/10.1080/14724040902953076
93. Zelenka, J. (2015). *Cestovní ruch. Marketing [Tourism. Marketing]* (p. 247). Gaudeamus Hradec Králové. ISBN 978-80-7435-543-1.
94. Budeanu, A., et al. (2016). Sustainable tourism, progress, challenges and opportunities: An introduction. *Journal of Cleaner Production, 111*(Part B), 285–294. https://doi.org/10.1016/j.jclepro.2015.10.027
95. Cohen, E. (1988a). Traditions in the qualitative sociology of tourism. *Annals of Tourism Research, 15*(1), 29–46. https://doi.org/10.1016/0160-7383(88)90069-2
96. UNESCO. (2015). Statutes and operational guidelines for UNESCO global geoparks, [online]. Retrieved August 11, 2022, from https://unesdoc.unesco.org/ark:/48223/pf0000234539

97. Llach, J., Marimon, F., & Alonso-Almeida, M. M. (2015). Social accountability 8000 standard certification: Analysis of worldwide diffusion. *Journal of Cleaner Production, 93*, 288–298. https://doi.org/10.1016/j.jclepro.2015.01.044
98. Ap, J., & Crompton, J. L. (1998). Developing and testing a tourism impact scale. *Journal of Travel Research, 37*(2) SAGE Publications, 120–130.
99. Cohen, E., & Cohen, S. A. (2012). Current sociological theories and issues in tourism. *Annals of Tourism Research, 39*(4), 2177–2202. https://doi.org/10.1016/j.annals.2012.07.009
100. Weaver, D. B. (2012b). Organic, incremental and induced paths to sustainable mass tourism convergence. *Tourism Management, 33*(5), 1030–1037. https://doi.org/10.1016/j.tourman.2011.08.011
101. Pásková, M., & Zelenka, J. (2018). Společensky odpovědný cestovní ruch. IDEA Servis Praha, 191 str., ISBN 978-80-85970-91-3.
102. Singapore Airlines. (2022). Sustainability at Singapore Airline, [online]. Retrieved August 4, 2022, from https://www.singaporeair.com/en_UK/sg/about-us/sustainability/
103. Hahn, H., & Kagelmann, H. J. (1993). *Tourismuspsychologie und Tourismussociologie: ein Handbuch zur Tourismuswissenschaft* (p. 630). Quintessenz.
104. Sin, H. L., & Minca, C. (2014). Touring responsibility: The trouble with 'going local' in community-based tourism in Thailand. *Geoforum, 51*, 96–106. https://doi.org/10.1016/j.geoforum.2013.10.004
105. UNWTO. (2004). Indicators of sustainable development for tourism destinations. A guidebook (English version), p. 507, ISBN 978-92-844-0726-2.
106. Han, H. (2021). Consumer behavior and environmental sustainability in tourism and hospitality: A review of theories, concepts, and latest research. *Journal of Sustainable Tourism, 29*(7), 1021–1042. https://doi.org/10.1080/09669582.2021.1903019
107. Pásková, M. (2018). Can indigenous knowledge contribute to the sustainability management of the Aspiring Rio Coco Geopark, Nicaragua? *Geosciences (Switzerland), 8*(8), 277. https://doi.org/10.3390/geosciences8080277
108. Rezidor. (2022). The Rezidor hotel group – Code of Ethics, [online]. Retrieved April 10, 2022, from https://www.rezidor.com/sv/about-us/our-code-business-ethics/
109. Gilbert, E. W., & Litt, B. (1939). The growth of Inland and seaside health resorts in England. *The Scottish Geographical Magazine, 55*(1), 16–35. https://doi.org/10.1080/00369223908735100
110. Christaller, W. (1963). Some considerations of tourism locations in Europe: The peripheral regions – Under-developed countries – Recreation areas. *Regional Science Association Papers, 12*, 95–105.
111. MHI. (2018). Code of ethics – Meliá Hotels International, [online]. Retrieved August 10, 2022, from https://www.meliahotelsinternational.com/en/shareholders-investors/corporate-governance/code-of-ethics
112. Luo, W., Tang, P., Jiang, L., & Su, M. M. (2020). Influencing mechanism of tourist social responsibility awareness on environmentally responsible behavior. *Journal of Cleaner Production, 271*, 122565. https://doi.org/10.1016/j.jclepro.2020.122565
113. Wang, N. (1999). Rethinking authenticity in tourism experience. *Annals of Tourism Research, 26*(2), 349–370. https://doi.org/10.1016/S0160-7383(98)00103-0
114. Wang, S., & Chen, J. S. (2015). The influence of place identity on perceived tourism impacts. *Annals of Tourism Research, 52*, 16–28. https://doi.org/10.1016/j.annals.2015.02.016
115. Garrod, B., & Fyall, A. (1998). Beyond the rhetoric of sustainable tourism? *Tourism Management, 19*(3), 199–212. https://doi.org/10.1016/S0261-5177(98)00013-2
116. Singh, A. K., Zhang, Y., & Anu. (2023). Understanding the evolution of environment, social and governance research: Novel implications from bibliometric and network analysis. *Evaluation Review, 47*(2), 350–386. https://doi.org/10.1177/0193841X221121244
117. Butler, R., & Hinch, T. (Eds.). (2007). *Tourism and indigenous peoples* (p. 380). Butterworth-Heinemann. ISBN 9780750664462.
118. Parsa, H. G., & Narapareddy, V. (2015). *Sustainability, social responsibility, and innovations in the hospitality industry* (p. 370). Apple Academic Press. ISBN 9781926895673.

119. Kim, M., & Kim, V. (2014). Corporate social responsibility and shareholder value of restaurant firms. *International Journal of Hospitality Management, 40*, 120–129. https://doi.org/10.1016/j.ijhm.2014.03.006
120. Carr, A., Ruhanen, L., & Whitford, M. (2016). Indigenous peoples and tourism: The challenges and opportunities for sustainable tourism. *Journal of Sustainable Tourism, 24*(8–9), 1067–1079. https://doi.org/10.1080/09669582.2016.1206112
121. Williams, P. W., & Ponsford, I. F. (2009). Confronting tourism's environmental paradox: Transitioning for sustainable tourism. *Futures, 41*(6), 396–404. https://doi.org/10.1016/j.futures.2008.11.019
122. Phillips, A. (2004). The new paradigm for protected areas, IUCN, [online]. Retrieved October 13, 2007, from www.iucn.org/themes/wcpa/events/previous/pavisitors.pdf
123. Tilikidou, I., Delistavrou, A., & Sapountzis, N. (2014). Customers' ethical behaviour towards hotels. *Procedia Economics and Finance, 9*, 425–432. https://doi.org/10.1016/S2212-5671(14)00044-6
124. Tol, R. S. J. (2007). The impact of a carbon tax on international tourism. *Transportation Research Part D: Transport and Environment, 12*(2), 129–142. https://doi.org/10.1016/j.trd.2007.01.004
125. Mihalic, T. (2016). Sustainable-responsible tourism discourse – Towards 'responsustable' tourism. *Journal of Cleaner Production, 111*(Part B), 461–470. https://doi.org/10.1016/j.jclepro.2014.12.062
126. McGehee, N. G. (2002). Alternative tourism and social movements. *Annals of Tourism Research, 29*(1), 124–143. https://doi.org/10.1016/S0160-7383(01)00027-5
127. Kim, S. (2021). CSR activities by theme parks in South Korea: What they do and report? *International Theme & Amusement Park Journal, 2*(3), 64–75.
128. Hu, C. (1996). Diverse developments in travel and tourism marketing: A thematic approach. *International Journal of Contemporary Hospitality Management, 8*(7), 33–43. https://doi.org/10.1108/09596119610152032
129. Young, A., & Kielkiewicz-Young, A. (2001). Sustainable supply network management. *Corporate Environmental Strategy, 8*(3), 260–268. https://doi.org/10.1016/S1066-7938(01)00122-1
130. Buhalis, D., & Law, R. (2008). Progress in information technology and tourism management: 20 years on and 10 years after the Internet – The state of eTourism research. *Tourism Management, 29*, 609–623. https://doi.org/10.1016/j.tourman.2008.01.005
131. Rosenow, J. E., & Pulsipher, G. L. (1979). *Tourism: The good, the bad and the ugly* (p. 264). Century Three Press.
132. Lee, J.-S., Hsu, L.-T., Han, H., & Kim, Y. (2010). Understanding how consumers view green hotels: How a hotel's green image can influence behavioural intentions. *Journal of Sustainable Tourism, 18*(7), 901–914. https://doi.org/10.1080/09669581003777747
133. Tosun, C. (2001). Challenges of sustainable tourism development in the developing world: The case of Turkey. *Tourism Management, 22*(3), 289–303. https://doi.org/10.1016/S0261-5177(00)00060-1
134. Heukemes, N. (2001). *Loving them to death? Sustainable tourism in Europe's nature and national parks* (p. 136). Europarc Federation.
135. Warnken, J., Bradley, M., & Guilding, C. (2005). Eco-resorts vs. mainstream accommodation providers: An investigation of the viability of benchmarking environmental performance. *Tourism Management, 26*(3), 367–379. https://doi.org/10.1016/j.tourman.2003.11.017
136. Sin, H. L. (2009). Volunteer tourism – "Involve me and I will learn"? *Annals of Tourism Research, 36*(3), 480–501. https://doi.org/10.1016/j.annals.2009.03.001
137. Vlachos, P. A., Tsamakos, A., Vrechopoulos, A. P., & Avramidis, P. K. (2009). Corporate social responsibility: Attributions, loyalty, and the mediating role of trust. *Journal of the Academy of Marketing Science, 37*(2), 170–180. https://doi.org/10.1007/s11747-008-0117-x
138. Cole, D. C. (2004). Monitoring and management of recreation in protected areas: The Contributions and limitations of science. *Environmental Management, 2*, 10–17.

139. Cole, D. N., & Monz, C. A. (2003). Impacts of camping on vegetation: Response and recovery following acute and chronic disturbance. *Environmental Management, 32*, 693–705. https://doi.org/10.1007/s00267-003-0046-x
140. Li, W. (2012). Study on the relationships between corporate social responsibility and corporate international competitiveness. *Energy Procedia, 17*(Part A), 567–572. https://doi.org/10.1016/j.egypro.2012.02.137
141. Pásková, M., & Dowling, R. K. (2014). The usage of local and indigenous knowledge in the management of geotourism destinations. In Conference Proceedings of 14th International Multidisciplinary Scientific Geoconference SGEM 2014, Geoconference on Ecology, Economics, Education and Legislation. Ecology and Environmental Protection. Albena, Bulgaria, Vol. II, pp. 805–812, ISBN 978-619-7105-18-6.
142. WCTE. (2022). The responsible tourist, world committee on tourism ethics, [online]. Retrieved October 4, 2022, from https://www.unwto.org/responsible-tourist
143. Wearing, S. (2001). *Volunteer tourism: Experiences that make a difference*. CAB International.
144. Bramwell, B. (2015). Theoretical activity in sustainable tourism research. *Annals of Tourism Research, 54*, 204–218. https://doi.org/10.1016/j.annals.2015.07.005
145. Kreisel, B. (2001). Generating income in protected areas. Issues of sustainable tourism development, EUROPARC: Loving them to death? Sustainable tourism in Europe's nature and national parks, Grafenau 2001, [online]. Retrieved July 5, 2023, from http://www.macaulay.ac.uk/workshop/bulgaria02/bettinakreisel_bulgaria02.pdf
146. ISO 13810:2022: Tourism and related services. Visits to industrial, natural, cultural and historical sites. Requirements and recommendations.
147. Cohen, E. (1988b). Authenticity and commoditization in tourism. *Annals of Tourism Research, 15*(3), 371–386. https://doi.org/10.1016/0160-7383(88)90028-X
148. MacCannell, D. (1999). *The tourist: A new theory of the leisure class* (p. 277). University of California Press, ISBN 9780520218925.
149. Deery, M., Jago, L., & Fredline, L. (2012). Rethinking social impacts of tourism research: A new research agenda. *Tourism Management, 33*(1), 64–73. https://doi.org/10.1016/j.tourman.2011.01.026
150. Eagles, P. J. F., McCool, S. F., & Haynes, C. D. (2002). Sustainable tourism in protected areas: Guidelines for planning and management. UNEP/IUCN (World Commission on Protected Areas)/WTO, Best Practice Protected Area Guidelines Series No. 8, p. 183, ISBN 2-8317-0648-3, [online]. Retrieved October 8, 2011, from www.uneptie.org/pc/tourism/library/sust_prot_areas.htm, www.iucn.org/dbtwwpd/-edocs/PAG-008.pdf
151. Martin, B. S., & Uysal, M. (1990). An examination of the relationship between carrying capacity and the tourism lifecycle: Management and policy implications. *Journal of Environmental Management, 31*, 327–333. https://doi.org/10.1016/S0301-4797(05)80061-1
152. Mazurkiewicz, P., & Grenna, L. (2003). Corporate social responsibility and multistakeholder dialogue. Towards environmental behavioral change. World Bank, p. 20, [online]. Retrieved January 22, 2024, from https://policycommons.net/artifacts/1442709/corporate-social-responsibility-and-multi-stakeholder-dialogue/2073571/
153. Milman, A., Okumus, F., & Dickson, D. (2010). The contribution of theme parks and attractions to the social and economic sustainability of destinations. *Worldwide Hospitality and Tourism Themes, 2*(3), 338–345. https://doi.org/10.1108/17554211011052249
154. Tsaur, S. H., Lin, Y.-C., & Lin, J.-H. (2006). Evaluating ecotourism sustainability from the integrated perspective of resource, community and tourism. *Tourism Management, 27*(4), 640–653. https://doi.org/10.1016/j.tourman.2005.02.006
155. Rodrigue, J. P., & Notteboom, T. (2013). The geography of cruises: Itineraries, not destinations. *Applied Geography, 38*, 31–42. https://doi.org/10.1016/j.apgeog.2012.11.011
156. Melnyk, S. A., Sroufe, R. P., & Calantone, R. (2003). Assessing the impact of environmental management systems on corporate and environmental performance. *Journal of Operations Management, 21*(3), 329–351. https://doi.org/10.1016/S0272-6963(02)00109-2
157. Smith, V. (Ed.). (1989). *Tourism: The sacred journey. Hosts and guests: The anthropology of tourism* (2nd ed. ISBN 9780812212808,, p. 352). University of Pennsylvania Press.

158. Cole, S. (2007). Beyond authenticity and commodification. *Annals of Tourism Research, 34*(4), 943–960. https://doi.org/10.1016/j.annals.2007.05.004
159. Hughes, G. (1995). Authenticity in tourism. *Annals of Tourism Research, 22*(4), 781–803. https://doi.org/10.1016/0160-7383(95)00020-X
160. Rahmawati, P. I., Jiang, M., Law, A., Wiranatha, A. S., & DeLacy, T. (2019). Spirituality and corporate social responsibility: An empirical narrative from the Balinese tourism industry. *Journal of Sustainable Tourism, 27*(1), 156–172. https://doi.org/10.1080/09669582.2018.1513006
161. Su, X., Wang, H., & Wen, T. (2013). Profit, responsibility, and the moral economy of tourism. *Annals of Tourism Research, 43*, 231–250. https://doi.org/10.1016/j.annals.2013.07.008
162. Vrontis, D., & Thrassou, A. (2006). Situation analysis and strategic planning: An empirical case study in the UK beverage industry. *Innovative Marketing, 2*(2), 134–151.
163. Sartori, I., & Hestnes, A. G. (2007). Energy use in the life cycle of conventional and low-energy buildings: A review article. *Energy and Buildings, 39*(3), 249–257. https://doi.org/10.1016/j.enbuild.2006.07.001
164. Vujko, A., Plavša, J., Petrović, M. D., Radovanović, M., & Gajić, T. (2017). Modelling of carrying capacity in National Park - Fruška Gora (Serbia) case study. *Open Geosci., 9*, 61–72. https://doi.org/10.1515/geo-2017-0005
165. Wells, V. K., et al. (2015). Heritage tourism, CSR and the role of employee environmental behaviour. *Tourism Management, 48*, 399–413. https://doi.org/10.1016/j.tourman.2014.12.015
166. Šlesinger, J., Kozielová, Z., & Najmanová, K. (2007). Čistší produkce. Příručka pro podniky a veřejnou správu [Cleaner Production. A guide for business and public administration]. CENIA, Prague, p. 100, [online]. Retrieved January 14, 2024, from https://invenio.nusl.cz/record/361669/files/nusl-361669_1.pdf
167. Dwyer, L., et al. (2009). Destination and enterprise management for a tourism future. *Tourism Management, 30*(1), 63–74. https://doi.org/10.1016/j.tourman.2008.04.002
168. Middleton, V. T. C. (1998). *Marketing in travel and tourism* (p. 393). Buttherworth Heinemann, ISBN 0-7506-0973-7.
169. Belnap, J., et al. (1997). The Visitor Experience and Resource Protection (VERP) framework: A handbook for planners and managers. U.S. Department of the Interior, [online]. Retrieved November, 2017, from https://www.fs.fed.us/cdt/carrying_capacity/verphandbook_1997.pdf.
170. Beunen, R., Regnerus, H. D., & Jaarsma, C. F. (2008). Gateways as a means of visitor management in national parks and protected areas. *Tourism Management, 29*(1), 138–145. https://doi.org/10.1016/j.tourman.2007.03.017
171. Tsai, W. H., Chang, Y.-C., Lin, S.-J., Chen, H.-C., & Chu, P.-Y. (2014). A green approach to the weight reduction of aircraft cabins. *Journal of Air Transport Management, 40*, 65–77. https://doi.org/10.1016/j.jairtraman.2014.06.004
172. Yunis. (2006). Making tourism more sustainable. Helpful instruments and practice examples in Europa. Prezentace 3. 2. 2006, Hannover, 19. s., [online]. Retrieved August 14, 2014, from http://www.ecotrans.org/docs/1_hamele_Yunis_intro_aims.pdf
173. Zelenka, J., & Kacetl, J. (2014). The concept of carrying capacity in tourism. *Amfiteatru Economic, 16*(36), 641–654.
174. Marsden, K., et al. (2014). Integration of biodiversity into CSR processes in tourism, p. 44, [online]. Retrieved January 14, 2024, from https://adelphi.de/system/files/mediathek/bilder/csrbiodiv-baseline_study-barrierfrei-en_final_0.pdf
175. COMMITTEE. (2002). Public participation in protected area management best practice. Parks and Wildlife Commission of the Northern Territory, [online]. Retrieved August 9, 2022, from https://www.environment.gov.au/system/files/resources/45209a20-1ab1-40d1-aa26-cf4838948782/files/public-participation.doc
176. CBD and UNEP. (2007a). Managing tourism & biodiversity. User's manual on the CBD guidelines on biodiversity and tourism development. CBD a UNEP, ISBN 92-9225-069-8, p. 125, [online]. Retrieved August 19, 2022, from http://tourism.cbd.int/manual.shtml

177. Beritelli, P. (2011). Cooperation among prominent actors in a tourist destination. *Annals of Tourism Research, 38*(2), 607–629. https://doi.org/10.1016/j.annals.2010.11.015
178. Hatipoglu, B., Alvarez, M. D., & Ertuna, B. (2016). Barriers to stakeholder involvement in the planning of sustainable tourism: The case of the Thrace region in Turkey. *Journal of Cleaner Production, 111*(Part B), 306–317. https://doi.org/10.1016/j.jclepro.2014.11.059
179. Haywood, K. M. (1988). Responsible and responsive tourism planning in the Community. *Tourism Management, 9*(2), 105–118. https://doi.org/10.1016/0261-5177(88)90020-9
180. Weiler, B., & Black, R. (2015). The changing face of the tour guide: One-way communicator to choreographer to co-creator of the tourist experience. *Tourism Recreation Research, 40*(3), 364–378. https://doi.org/10.1080/02508281.2015.1083742
181. Jamal, T., & Camargo, B. A. (2014). Sustainable tourism, justice and an ethic of care: Toward the Just Destination. *Journal of Sustainable Tourism, 22*(1), 11–30. https://doi.org/10.1080/09669582.2013.786084
182. Namkung, Y., & Jang, S. (2013). Effects of restaurant green practices on brand equity formation: Do green practices really matter? *International Journal of Hospitality Management, 33*, 85–95. https://doi.org/10.1016/j.ijhm.2012.06.006
183. Pásková, M. (2012). Environmentalistika cestovního ruchu [Environmental tourism]. *Czech Journal of Tourism, 1*(2), 77–113.
184. Pásková, M. (1999). Negative impacts of tourism and sustainable tourism as an alternative for regional development. In P. Hlavinková, & J. Munzar (Eds.), *Regional prosperity and sustainability. Proceedings of the 3rd Moravian Geographical Conference CONGEO ′99* (pp. 155–167). Slavkov u Brna.
185. Gössling, S., et al. (2005). The eco-efficiency of tourism. *Ecological Economics, 54*(4), 417–434. https://doi.org/10.1016/j.ecolecon.2004.10.006
186. Gössling, S., et al. (2011). Food management in tourism: Reducing tourism's carbon 'foodprint'. *Tourism Management, 32*(3), 534–543. https://doi.org/10.1016/j.tourman.2010.04.006
187. Gössling, S., & Buckley, R. (2016). Carbon labels in tourism: Persuasive communication? *Journal of Cleaner Production, 111*(Part B), 358–369. https://doi.org/10.1016/j.jclepro.2014.08.067
188. Gössling, S., Scott, D., & Hall, C. M. (2013). Challenges of tourism in a low-carbon economy. *WIREs Climate Change, 4*, 525–538. https://doi.org/10.1002/wcc.243
189. Hunter, C. (1997). Sustainable tourism as an adaptive paradigm. *Annals of Tourism Research, 24*(4), 850–867. https://doi.org/10.1016/S0160-7383(97)00036-4
190. Farrell, B. H., & Twining-Ward, L. (2004). Reconceptualizing tourism. *Annals of Tourism Research, 31*(2), 274–295. https://doi.org/10.1016/j.annals.2003.12.002
191. Li, T.-T., Wang, K., Sueyoshi, T., & Wang, D. D. (2021). ESG: Research progress and future prospects. *Sustainability, 13*(21), 11663. https://doi.org/10.3390/su132111663
192. Boluk, K. A., Cavaliere, C. T., & Higgins-Desbiolles, F. (2019). A critical framework for interrogating the United Nations Sustainable Development Goals 2030 Agenda in tourism. *Journal of Sustainable Tourism, 27*(7), 847–864. https://doi.org/10.1080/09669582.2019.1619748
193. Bramwell, B., Higham, J., Lane, B., & Miller, G. (2017). Twenty-five years of sustainable tourism and the Journal of Sustainable Tourism: Looking back and moving forward. *Journal of Sustainable Tourism, 25*(1), 1–9. https://doi.org/10.1080/09669582.2017.1251689
194. Hall, C. M. (2019). Constructing sustainable tourism development: The 2030 agenda and the managerial ecology of sustainable tourism. *Journal of Sustainable Tourism, 27*(7), 1044–1060. https://doi.org/10.1080/09669582.2018.1560456
195. Pásková, M. (2002). Destination life cycle of the historic town Český Krumlov. *Tourism. An International Interdisciplinary Journal. Special Issue Cultural Tourism, 50*(3), 249–264.
196. EUROPARC Federation. (2010). European Charter for sustainable tourism in protected areas, EUROPARC Federation, [online]. Retrieved August 9, 2022, from http://www.europarc.org/wp-content/uploads/2015/05/2010-European-Charter-for-Sustainable-Tourism-in-Protected-Areas.pdf

197. Millar, M., & Baloglu, S. (2011). Hotel guests' preferences for green guest room attributes. *Cornell Hospitality Quarterly, 52*(3), 302–311. https://doi.org/10.1177/1938965511409031
198. Seo, K., Moon, J., & Lee, S. (2015). Synergy of corporate social responsibility and service quality for airlines: The moderating role of carrier type. *Journal of Air Transport Management, 47*, 126–134. https://doi.org/10.1016/j.jairtraman.2015.05.011
199. Christ, C., Hillel, O., Matus, S., & Sweeting, J. (2003). Tourism and biodiversity: Mapping tourism's global footprint. UNEP a Conservation International, p. 70, [online]. Retrieved November 14, 2016, from http://www.unep.org/publications/search/pub_details_s.asp?ID=3788
200. Eagles, P. F. J. (2009). Governance of recreation and tourism partnerships in parks and protected areas. *Journal of Sustainable Tourism, 17*(2), 231–248. https://doi.org/10.1080/09669580802495725
201. Kindler, E. (2016). A comparison of the concepts: Ecosystem services and forest functions to improve interdisciplinary exchange. *Forest Policy and Economics, 67*, 52–59. https://doi.org/10.1016/j.forpol.2016.03.011
202. Naess, A. (2009). *Ecology, community and lifestyle. Outline of ecosophy* (p. 223). Cambridge University Press. https://doi.org/10.1017/CBO9780511525599
203. Skudiene, V., & Auruskeviciene, V. (2012). The contribution of corporate social responsibility to internal employee motivation. *Baltic Journal of Management, 7*(1), 49–67. https://doi.org/10.1108/17465261211197421
204. Saeidi, S. P., Sofian, S., Saeidi, P., Saeidi, S. P., & Saaeidi, S. A. (2015). How does corporate social responsibility contribute to firm financial performance? The mediating role of competitive advantage, reputation, and customer satisfaction. *Journal of Business Research, 68*(2), 341–350. https://doi.org/10.1016/j.jbusres.2014.06.024
205. TIAC. (2022). Code of ethics and guidelines for sustainable tourism, tourism ndustry Association of Canada and National Round Table on the Environment and the Economy, [online]. Retrieved August 14, 2022, from http://nrt-trn.ca/wp-content/uploads/2011/08/sustainable-tourism.pdf
206. Holden, A. (2003). In need of new environmental ethics for tourism? *Annals of Tourism Research, 30*(1), 94–108. https://doi.org/10.1016/S0160-7383(02)00030-0
207. Holden, A. (2009). The environment-tourism nexus: Influence of market ethics. *Annals of Tourism Research, 36*(3), 373–389. https://doi.org/10.1016/j.annals.2008.10.009
208. Ryan, C. (1991). *Recreation tourism – A social science perspective* (p. 227). Routledge.
209. UNEP. (2003). Tourism and local Agenda 21. The role of local authorities in sustainable tourism, p. 59, ISBN 92-807-2267-0.
210. Stronza, A., & Gordillo, J. (2008). Community views of ecotourism. *Annals of Tourism Research, 35*(2), 448–468. https://doi.org/10.1016/j.annals.2008.01.002
211. McGehee, N. G., & Santos, C. A. (2005). Social change, discourse and volunteer tourism. *Annals of Tourism Research, 32*(3), 760–779. https://doi.org/10.1016/j.annals.2004.12.002
212. Russo, A. P. (2002a). The "vicious circle" of tourism development in heritage cities. *Annals of Tourism Research, 29*(1), 165–182. https://doi.org/10.1016/S0160-7383(01)00029-9
213. Font, X., et al. (2012). Corporate social responsibility: The disclosure–performance gap. *Tourism Management, 33*(6), 1544–1553. https://doi.org/10.1016/j.tourman.2012.02.012
214. Garay, L., & Font, X. (2012). Doing good to do well? Corporate social responsibility reasons, practices and impacts in small and medium accommodation enterprises. *International Journal of Hospitality Management, 31*(2), 329–337. https://doi.org/10.1016/j.ijhm.2011.04.013
215. Benavides-Velasco, C. A., Quintana-García, C., & Marchante-Lara, M. (2014). Total quality management, corporate social responsibility and performance in the hotel industry. *International Journal of Hospitality Management, 41*, 77–87. https://doi.org/10.1016/j.ijhm.2014.05.003

Chapter 3
Current Understanding of Social Responsibility and Basic Knowledge on Its Application

Two terms, in particular, are used in studies of responsibility, voluntarily taken by organisations, namely corporate social responsibility (CSR, sometimes also referred to as corporate responsibility) and Corporate Social and Environmental Responsibility (CSER). In this chapter, the more general term of **social responsibility** is used for the tourism application discussed below, which encompasses the economic, socio-cultural and environmental responsibilities of organisations and individuals at different geographical levels (local, regional, global). The long-established term CSR will also be used when discussing social responsibility implemented by businesses or institutions.

3.1 Current Understanding of Corporate Social Responsibility

CSR is defined differently by different experts, see, for example the overview of different definitions ([1]:2–3; [2]:4–5; [3]:8). For the purpose of this publication, which emphasises the links with the environment of the implementer, a modified concept and supplemented according to ISO 20121 ([4]:337) was chosen. Corporate social responsibility is a continuous, long-term and planned process of managing organisational and personal values, based on the vision of the organisation. It concerns people (the various actors in relation to the social responsibility implementer), the environment and organisational policies and is influenced by political interests. The adoption of CSR is often associated with financial and other benefits for its implementer, as confirmed especially for long-term oriented financial benefits by Lin, Yang and Liu [5] based on their comparative analysis of many studies and their own research. Its adoption also brings about desirable fundamental changes in the environmentally oriented behaviour of organisations ([3]). Properly applied CSR in business can represent a competitive advantage, as documented for the tourism

M. Pásková, J. Zelenka, *Social Responsibility in Tourism*, Tourism on the Verge,
https://doi.org/10.1007/978-3-031-61609-9_3

industry by Arnaudov and Koseska [6]). They attribute this increased competitiveness not only to strict compliance with relevant legislation, but above all to a conscious approach, striving for a balanced pursuit of interrelated economic, environmental and social objectives. They consider this holistic approach to be beneficial for both the local residents and environment as well as society as a whole.

The nature of tourism is specific. In addition to the activities of companies that supply tourism market with their services, other important components play a role. These include individual service providers or their groups (e.g. private accommodation facilities), the implementation of destination management and visitor management, distinct from the management of organisations and the activities of individuals, especially tourism participants. For this reason, the more general term "social responsibility" is predominantly used in this publication. This term encompasses a less formal notion of the responsibility of an individual or a group of individuals, often based on the personal values of these individuals, their knowledge, their identification with a particular pattern or rules of behaviour and with a particular place (local identity), the availability of correct information, their ethics and the size of their environmental bubble ([7]:167). There is also an obvious link with the empathy of these individuals in relation to other tourism actors (e.g. the local community).

Tourism and other human activities are gradually moving into space (remote exploration of the Earth, satellite navigation, space tourism etc.). Therefore, it is very timely to extend social responsibility from Earth to the surrounding and gradually more distant space.[1]

3.2 Basic Aspects of Social Responsibility

The basic aspects of social responsibility can be viewed from several different perspectives, as briefly outlined in this chapter, both in general and in relation to its application in tourism. The concept of social responsibility has undergone considerable development over the last 50 years in relation to the content of social responsibility in international business as well as its conceptualisation ([9]). Young and Kielkiewicz-Young [10] showed that at the turn of the millennium, in North America and Europe, supply chains were dominated by the implementation of environmentally friendly activities over the implementation of social responsibility activities. This, they argue, was due to the pressure of external factors (mainly legislation, market competitiveness and media attention) in relation to the environment and, conversely, the absence of pressure from external moral and ethical requirements in relation to society.

In the tourism industry, in addition to partial socio-cultural responsibility in relation to local communities, the emphasis is generally placed on the environmental

[1] See, e.g. considerations on a code of ethics for space exploration and development, according to [8].

dimension of social responsibility. Tzschentke, Kirk and Lynch [11] indicate important factors in the motivation of environmentally friendly social responsibility of hospitality service operators. Namely, the sense of enhancing the value of their own services, which is mainly influenced by the development of environmental awareness of management and employees, as well as personal, socio-cultural factors and specific local conditions.

3.2.1 Three Basic Dimensions of Social Responsibility

The three basic dimensions of social responsibility—environmental, socio-cultural and economic—including their typical components are illustrated in the diagrams in Figs. 3.1, 3.2, 3.3, 3.4, 3.5 and 3.6. These generally valid diagrams, expressing the main themes and focus of social responsibility, are fully applicable to tourism. At the same time, given the nature of tourism and the local conditions of destinations, their geographical characteristics, historical development, natural and cultural heritage and the nature and status of the local community, it is important to emphasise destination-specific aspects of social responsibility.

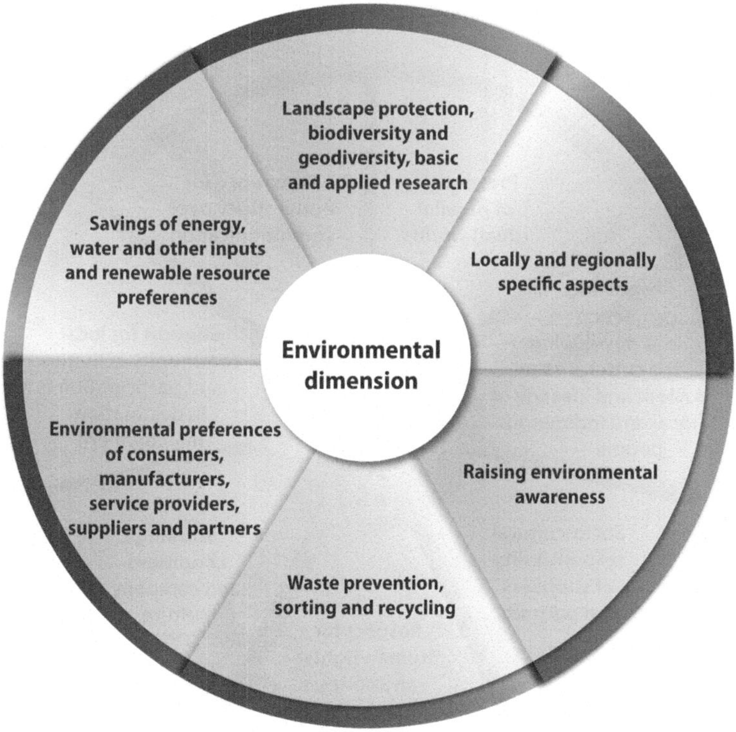

Fig. 3.1 The environmental dimension of social responsibility. Source: Inspired by [12]:66–78

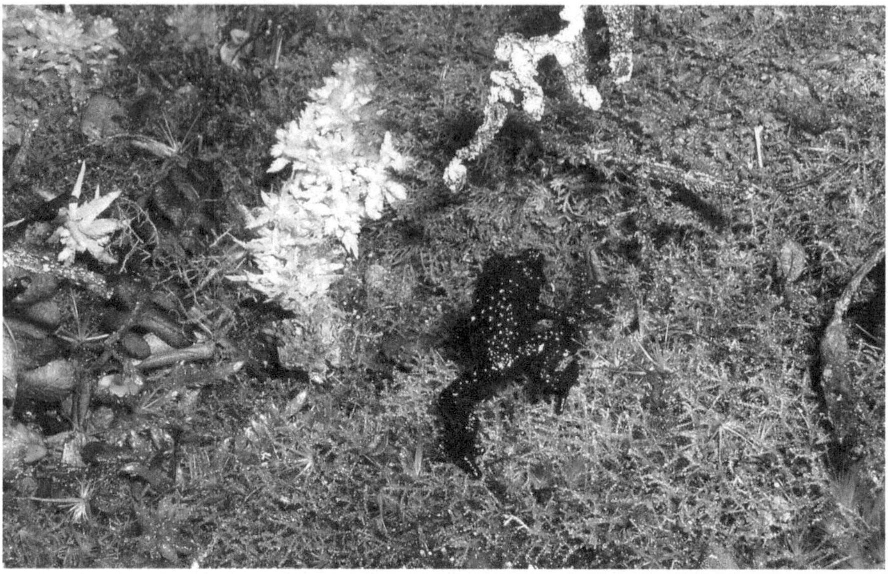

Fig. 3.2 Visitor regulations, both written and unwritten (and passed on by guides), are important for the conservation of endemic species such as this black frog *Anomaloglossus roraima*, living on Roraima Table Mountain, Canaima NP, the Venezuela–Brazil–Guyana tri-border area. Photo: J. Zelenka, 2006

Fig. 3.3 Socio-cultural dimensions of social responsibility in tourism. Source: Inspired by [12, 13, 14, 15]

Fig. 3.4 The traditional way of life, including local rituals and the traditional dress of the Indigenous inhabitants, is an important part of Easter Island's attraction for visitors. Respect for the institutions, appearance and traditional practices of the Rapanui places great demands on the social responsibility and sensitivity of all local tourism actors. Hanga Roa, Easter Island. Photo: M. Pásková, 2007

Fig. 3.5 The Miao Indigenous Peoples of the local communities continue to practice their traditional rituals of the god of the mountain. The part of the ceremony is offering of pig heads, peaches and oranges to the gods, as well as dancing and ritual drumming. Xiangxi UNESCO Global Geopark, China. Photo: M. Pásková, 2023

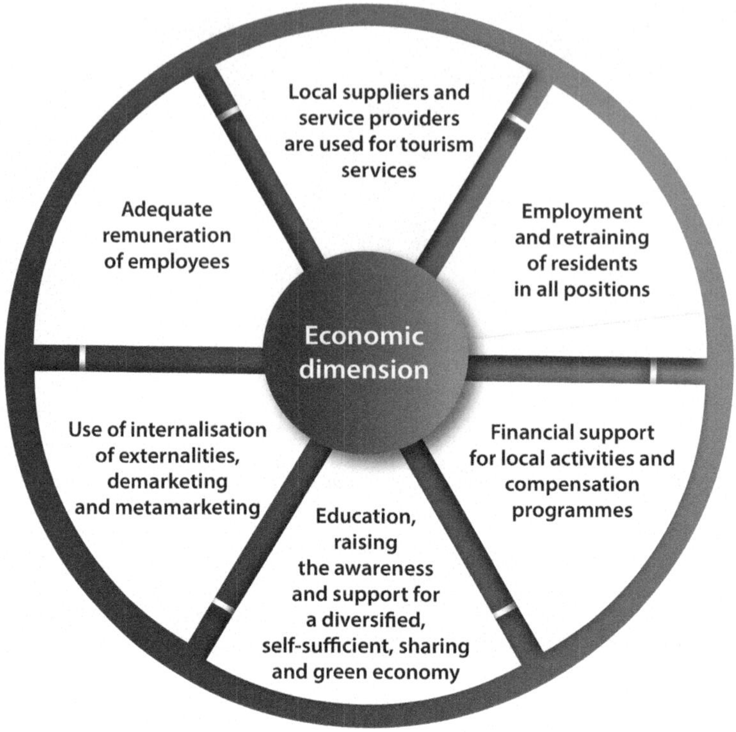

Fig. 3.6 Economic dimension of social responsibility in tourism. Source: Inspired by Černec [12]:62–66)

Environmental Dimensions of Social Responsibility in Tourism

The environmental dimension of social responsibility in tourism is most often explored and implemented in relation to visitors and tourism service providers. This is due to the key role of these tourism actors in optimising the impacts of tourism on the natural environment.[2] However, it should be explored and implemented in relation to all tourism actors, especially the local community (which can activate a higher level of social responsibility, e.g. through the interpretation of local natural heritage, a typical example being geoparks, see Sect. 6.1) and nature and landscape conservationists, in seeking a balance between protection and access to natural heritage. Appropriate access to and interpretation of natural heritage can create a relationship between visitors and nature and thus strengthen the environmental dimension of tourism actors' social responsibility.

Concerning the perception of social responsibility, some relevant studies (e.g. [17]) indicate that the environmental dimension is especially influential. The

[2] See [14, 16] for a more detailed discussion of the concept of optimising tourism impacts.

specific features of the environmental dimension of social responsibility in tourism, expressed in more general terms in the diagram in Fig. 3.1, are illustrated in the following case studies.

Case Study 1 Mountain Areas
Mountain areas are characterised by the threat of erosion, the presence of highly sensitive ecosystems to external influences (including human impact), valuable landscape character and a high interest in tourism development. It is therefore necessary to place great emphasis on developing the responsibility of tourism stakeholders for the conservation of the landscape and local ecosystems, and on developing tourism infrastructure in such a way as to reduce energy and water consumption (snowmaking) in addition to preventing erosion (typically on mountain trails). Responsible accommodation providers and visitors should prefer private accommodation to hotels. Visitors should also choose activities that are less aggressive towards the local ecosystems and landscape, e.g. by preferring healthier cross-country skiing to downhill skiing. Many mountain areas can be overloaded with tourism. The Krkonoše Mountains, Czechia, are a typical example in terms of the historical development of their utilisation. Destination management and nature and landscape conservationists need to take responsibility and implement visitor management in mountain areas with consistent management of visitor flows and activities (see, e.g. [18]).

Case Study 2 Unique Ecosystems with Endemic Species
There are several differently sized areas in the world with unique ecosystems whose specific characteristics determine the occurrence of endemic plant and animal species. These unique landscapes are gaining increasing popularity as tourist attractions. Model examples include the Galapagos Archipelago ([19]), the Indonesian island of Komodo, Antarctica ([20]) and the Table Mountains in the Guiana Highlands of Venezuela, with their strong restrictions for tourism development. The most frequently visited table mountain in Venezuela is Roraima (Fig. 3.2). Responsibility for its unique nature is realized by strict rules of visitor behaviour overseen by indigenous guides (of ethnic Pemon), the absence of visitor infrastructure, and daily visitor limits (designated natural shelters for overnight stays). The absence of tourism infrastructure and the accessibility of Roraima only after a 2-day strenuous trek through complicated mountainous terrain, including crossing the rivers, acts as a selection bias, preferring more responsible visitors with a higher level of respect for the environment of this pristine area and with little environmental bubble [14].

Case Study 3 Island Destinations

In island destinations, and most notably in those designated as Small Island Developing States (SIDS), tourism plays an important role not only in economic and social development, but also in the conservation of nature and care for cultural heritage. Typical problems in these destinations include water scarcity, the production of large amounts of waste coupled with minimal waste management, threats to local endemic species and the lack of energy sources. Easter Island is a typical example, with its high dependence on tourism, its waste disposal site near the only source of drinking water at the Rano Kau crater, and its incipient efforts to harness renewable energy by building wind farms on the Poike Peninsula. On small islands, therefore, great emphasis must be placed on responsible and planned management of water, waste (especially waste prevention), land-use, landscape planning, nature conservation and the utilisation of alternative resources. It is also important to raise the environmental awareness of all tourism actors, which should significantly increase the environmental responsibility of tourism service providers, their suppliers and partners. Prevention is key, with an emphasis on preserving landscape connectivity, protecting biodiversity by preventing the penetration of non-native and especially invasive species of flora and fauna, and preventing noise and light pollution as well as landscape fragmentation. Social responsibility was already comprehensively discussed in the final report of the International Conference on Sustainable Tourism in SIDS and other Islands, held in Lanzarote, Spain, in 1998 [21].

Socio-Cultural Dimensions of Social Responsibility in Tourism

The socio-cultural dimension of social responsibility in tourism is most often studied and implemented in relation to the local community in order to prevent and compensate for its adverse effects on its culture, lifestyle and standard of living (here it is intertwined with the economic dimension). However, it should be examined and implemented in relation to all tourism actors, especially visitors. As for the latter, socio-cultural responsibility can be initiated and strengthened through an appropriate interpretation of the local cultural heritage (here linked to its appropriate preventive protection), the local culture and ways of thinking. As a result, the visitors' social responsibility may deepen through a varying degree of acceptance of different ways of thinking, including a move away from an emphasis on consumer behaviour and material values, as well as a broader spiritual outlook and a deeper experience. The socio-cultural dimension of social responsibility should also be explored and implemented in relation to the employees of tourism service providers, who play an important role in implementing social responsibility in a destination. They should be provided with decent and healthy working conditions and opportunities for skills development.

The specific features of the socio-cultural dimension of social responsibility in tourism are summarised in Fig. 3.3 and illustrated in Case Studies 4 and 5. These include the ever-present threat of the commercialisation of cultural heritage, especially intangible heritage in the form of the staging of traditions and local lifestyles, the folklorisation of ceremonies and festivals and the commodification of souvenirs (degradation of traditional know-how). Preventing these undesirable effects of tourism requires social responsibility on the part of visitors, whose demand for a unique experience, supported by their purchasing power, puts pressure on the commercialisation of the aforementioned collective intangible heritage. Social events and traditions, often determined by natural cycles, must be respected, but it is equally important to be empathetic to the degree of their intimacy and sacredness for the local community. On the part of local and Indigenous peoples, who have no experience of these unintended impacts of tourism, this requires a cultural self-awareness and identity strong enough to prevent them from wholeheartedly incorporating this heritage ("family silver") into the tourism market in the form of producing "unique" and "authentic" experiences for visitors according to their expectations and based on their realized purchasing power (Fig. 3.4). The point is to avoid the unintended substitution of cultural capital ([22]) by economic capital and the emergence of distinct dual worlds ([14]).

Case Study 4 Cultural Sites Designated as World Heritage Sites
As a World Heritage Site, Český Krumlov is the second most visited cultural tourism destination in Czechia, after Prague. The uniqueness of its castle grounds and the coherent historical centre has been preserved despite the many negative effects of staging or even certain manifestations of "cultural prostitution" on the appearance of the urban core and the cultural identity of the local population. This is to the credit of the post-1989 municipal leaders as well as other tourism actors, especially preservationists. After the Velvet revolution in the Czech Republic in 1989, they promoted strategic planning and sustainable development of tourism, with an emphasis on reducing visual pollution and other undesirable externalities of tourism, including the development of appropriate tourism infrastructure (guesthouses, restaurants, private accommodation; see [23]) that fits into the historic environment of the place. This long-term responsible approach results in the relatively low values of the irritation index of local residents, especially outside the historical centre and among the younger generation of residents, and in the strengthening of local identity and pride in the town, which are prerequisites for further responsible approach to tourism among local residents and other tourism actors.

Case Study 5 Indigenous People

A typical problem of tourism destinations, especially those both geographically and culturally isolated, is usually the adverse effects of tourism on the cultural identity, environment and quality of life of Indigenous peoples. Visitors are interested in the cultural distinctiveness of Indigenous communities. Still, they do not always respect local customs or sacred sites. Due to potential and realised visitor demand, there is pressure to change local lifestyles, which can lead to the staging of some exotic traditions. Isolated (especially island) destinations are therefore, as part of tourism development, at risk of turning attractive parts of the territory into "living open-air museums", folklorising traditions, reducing the diversity of the economy (in extreme cases to tourism monoculture) and losing traditional ways of life linked to a knowledge system built up by generations of indigenous peoples. An example of this is Easter Island (Rapa Nui), the world's most geographically isolated tourist destination, which, despite the efforts of the indigenous people Rapanui to maintain their cultural identity (Fig. 3.4), is threatened by the staging and folklorisation of their cultural identity, not least in the context of their February festival, Tapati Rapa Nui. Given the limited size of the island (163 sq km) and its dramatic environmental history, Easter Island is in danger of exceeding not only the socio-cultural, but also the physical, psychological and ecological dimensions of the carrying capacity for tourism development.

The traditional knowledge of plant and animal species and their habitats, natural processes, the specific concept and practical use of cosmology in the context of natural healing, construction and agriculture, the shared perception of nature as Mother Earth, and the language or dialect of Indigenous peoples reflected in the local toponymy—all of this in its totality forms a very valuable and vulnerable system of knowledge of natural ethnics and Indigenous communities in general (see also [24]). This knowledge represents an important potential for the implementation of social responsibility in tourism by destination management organisations. Traditional local and Indigenous knowledge should be respected in the preparation of the destination product, with care for destination resources and participatory destination management in general. For visitors, this type of intangible heritage offers a significant opportunity to enrich their travel experience and life inspiration. The more responsibility they demonstrate by their readiness to appreciate the distinct culture of Indigenous cultures and the more they respect local institutions, customs, etiquette, living or dining cultures and religious rituals (Fig. 3.5), the deeper the experience they gain from their travels.

The application of Indigenous knowledge in the process of protecting the natural and/or cultural heritage of a destination is often more effective than the application of conventional measures by the nature and landscape conservation authorities and/or by the built monument care authorities ([24]). As an example, Borrini-Feyerabend, Kothari and Oviedo [25]:86 cite the efforts of the Indian authorities to protect the precious ecosystems of Keoladeo National Park[3] by fencing off its territory and preventing pastoral activities of Indigenous peoples. They point out that, after many conflicts and protests by the Indigenous population, a very costly scientific study was carried out which proved that cattle grazing is, on the contrary, essential for the preservation of the ecological characteristics of this wetland habitat. They appreciate that regulated grazing, participatory management and a system of sharing local tourism revenues with the Indigenous population have subsequently been introduced.

Economic Dimension of Social Responsibility

Tourism holds significant economic potential, especially in developed countries transitioning into information and knowledge societies with a focus on leisure. However, it also faces challenges like economic leakage, where revenues and taxes often leave the destination. Moreover, the tourism industry exploits seasonal labour that may migrate to the destination on a seasonal basis. Not only the headquarters of some tourism businesses but also their suppliers are often located outside the destination. In many destinations, this creates a situation in which negative impacts outweigh positive economic benefits of tourism. The former includes disproportionately increasing levels of local resource sharing with visitors, financial leakage outside the destination, and negative externalities of tourism including economic ones (dualisation of society, marginalisation of local people, destination core inflation—see [14] for details). The latter include, in particular, raising the standard of living of local residents, the multiplier effect of tourism and improving infrastructure, which positively affect other areas of business. To address economic responsibility in tourism (see also ISO 26000 in Sect. 4.5 and Fair Trade approaches in Sect. 4.7) actions such as preferring local sourcing, promoting social responsibility amongst suppliers, fair employee compensation and financial support for local initiatives are crucial (see also ISO 26000 in Sect. 4.5 and Fair Trade approaches in Sect. 4.7). This should include employing residents, sourcing products from local suppliers and service providers, in spreading social responsibility in networks of suppliers and cooperating partners in horizontal and vertical cooperation, in appropriate remunerating of employees (see also SA8000 in Sect. 4.6). In this way, economically responsible tourism development should lead to a diversified and endogenous local economy of the given destination. Economic responsibility can also be carried out by financially supporting compensation programmes that take place both inside and outside the destination. An overview of the economic dimension of social responsibility in tourism is shown in Fig. 3.6.

[3] A UNESCO World Heritage Site and a wetland of international importance, declared under the Ramsar Convention.

3.2.2 Change of Product Quality in Tourism

The implementation of different approaches to social responsibility in tourism leads to changes in the quality of the tourism product at different levels—objectively measurable physical characteristics of the product,[4] perceived characteristics of the product (e.g. perception of the way the landscape is interpreted in geotourism, see Sects. 5.5 and 6.1) and "moral" characteristics of the product, including the degree of application of social responsibility.

For the physical characteristics of the product, the implementation of social responsibility can increase product quality. An example is the insulation of a hotel building, which leads to an increase in the thermal comfort of the guests at all times of the year. Nonetheless, it can also reduce the quality of the physical characteristics of the product—e.g. in ecotourism, modestly furnished accommodation made of local materials offers only a basic range of services. Similarly, the perceived product quality increases or decreases, although it may not directly correlate with the "objective" physical quality of the product offered. This is affected by the size of the environmental bubble and the associated value system of visitors. However, if the way social responsibility is implemented is in line with the declared social responsibility practices of the institution, there is always an increase in the "moral" quality of the product. This in turn relates to influencing the preferences and behaviour of visitors, their level of satisfaction with the product, their willingness to pay for increased product quality (Sect. 3.2.3) and their preferences for choosing the product. Therefore, the success of promoting social responsibility in tourism will depend to a large degree on extending the perceived quality to include the environmental and moral dimensions. This transformation of the concept of quality will be particularly important for tourism, since the people living in the tourism destination are part of its products. Quality perceived in this way is an expression of the ability to meet the needs of the consumer, the whole community and the environment. It therefore implies "self-transcendence"—not only taking responsibility for oneself but also taking social responsibility for oneself and society.

3.2.3 Willingness to Pay for Increased Product Quality by Implementing Social Responsibility

The institution can reflect the cost of implementing social responsibility in the price of the product. This increase can, for example be relatively high for declared eco-accommodation and ecotourism services. This may result in abuse or misuse of the social responsibility of visitors by a commercial entity. Ideally, the willingness to pay a higher price for an environmentally and/or morally superior tourism product, for which social responsibility is applied, is balanced by an increase in the product value.

[4] For example, locally differentiated lighting of hotel spaces with energy-saving appliances.

Kim and Han [26] studied the willingness to pay the same price for accommodation in environmentally friendly hotels as for accommodation in conventional hotels, despite moderate discomfort (e.g. longer use of bed linen and towels and use of recycled products). They used a modified model of the theory of planned behaviour ([26]:1001) to interpret the results. In doing so, they tested four hypotheses (modified, abbreviated, see Box 3.1), also established based on other studies (e.g. [27, 28, 29]).

Box 3.1 Hypotheses about Accommodation Customers. Source: [26]

H1 The customer's beliefs about the consequences of his/her behaviour have a positive effect on his attitude towards his/her own behaviour.

H2 Normative beliefs (other people's opinion of the person) have a positive effect on the customer's subjective norms.

H3 The customer's awareness of control options has a positive effect on his/her perceived ability to control behaviour.

H4 The customer's environmental concerns, his/her environmentally friendly behaviour, efficiency perceived by the customer, the customer's subjective norms and his/her perceived ability to control behaviour have a positive effect on the customer's intention to pay a price comparable to that of a conventional hotel in an environmentally friendly hotel.

However, their research did not support the hypothesis that customers' awareness of the importance of environmental issues and the perceived environmental effectiveness of his/her choice of such a hotel would have a positive effect on the environmental friendliness of their behaviour.

As other studies confirm (e.g. [30, 31]—see Sect. 3.3.7), the client's willingness to pay extra for the moral dimension of product quality can be expected to positively correlate with his/her environmental preferences (as illustrated by Han et al. [27]), and also with the extent to which the client will acknowledge his/her own share of the harm incurred in the consumption of a particular tourism product to the local community by reducing the environmental or social value of its habitat. As far as the external costs are concerned, it is to this extent that they should be internalized (reflected in the product price). Their accurate expression in monetary terms is very difficult and often impossible, not least because of their partly intangible nature, multiplier and/or synergy effects and accumulation processes. Thus, the customer pays to the extent that he/she feels responsible, e.g. when paying for CO_2 emissions during airplane flights, see the discussion on willingness to pay offsets for them ([32]).

3.2.4 Environmental Orientation and Environmental Behaviour of Tourism Participants

Environmental orientation, or the environmental responsibility of visitors, is a frequent topic of research. In particular, its degree and factors affecting its assumed correlation with the environmentally declared as well as actual behaviour of tourism participants, including the aforementioned (Sect. 3.2.3) willingness to pay for (moral) product quality, are examined. These assumptions, often also presented in the form of various behavioural models, are confirmed by a number of studies (e.g. [33, 34]). Selected studies and their results are commented on in more detail below.

For hotel guests, Huang et al. [34] confirmed the expected positive correlation between the degree of environmental orientation and environmentally friendly actual behaviour (e.g. longer use of bed linen and towels). According to them, more environmentally friendly behaviour was exhibited by women and younger guests. This behaviour is in line with another expected assumption confirmed by Han et al. [27] and Chen [35]—environmentally friendly hotels are preferred especially by guests with eco-friendly attitudes.

Environmentally responsible behaviour of tourism participants has been defined and its measurement method described by, e.g. Lee, Jan and Yang [36] and Pásková [14]. In order to predict it, various models have been used, linking environmentally oriented goals (e.g. environmental awareness, perceived effectiveness of activities and eco-reputation) and environmental behaviour of tourism participants.[5]

3.2.5 Social Responsibility and Externalities in Tourism

Tourism does not only cause negative ([39, 40, 41, 42]) but also positive market externalities. Positive market externalities of tourism include, for example intensified care for cultural monuments or environmentally friendly innovations, paid for with tourism revenues ([43]). The social responsibility of tourism actors should be related to the optimisation of market externalities, i.e. to the prevention or minimisation of negative externalities and the simultaneous maximisation of positive market externalities (Fig. 3.7). As Pásková and Zelenka [43] state, the implementation of social responsibility naturally optimises tourism effects, and thus leads to a more effective effort dedicated to increasing the sustainability of tourism. According to them, tourism actors practicing social responsibility have an opportunity to reduce their contribution to the negative environmental, economic and social impacts of tourism and increase their contribution to the benefits of tourism for local residents and their environment.

---

[5] See, e.g. Han and Yoon [37], Han [38] and the modified model of the theory of planned behaviour already mentioned in Sect. 3.2.2.

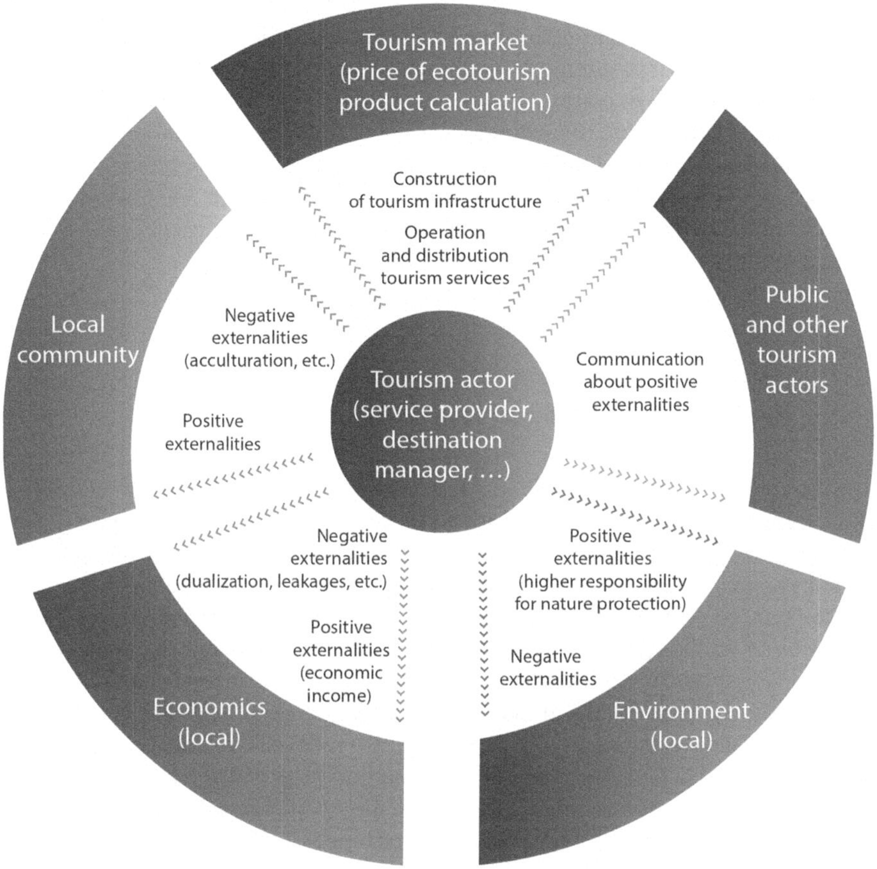

Fig. 3.7 Social responsibility and externalities in tourism. Source: [43]:783, graphically modified

Pásková and Zelenka [43] propose to incorporate related theoretical concepts into the study, research and practical application of the concept of social responsibility of tourism actors: social exchange theory ([7, 41]), the concept of carrying capacity ([41, 44, 45, 46, 47, 48, 49]), and the destination life cycle concept ([14, 41, 47, 49, 50, 51]).

3.3 Basic Knowledge on the Application of Social Responsibility

The following chapters provide a summary of findings regarding the application of social responsibility within a psychological framework, particularly concerning the human value system (Sect. 3.3.1). Additionally, the implementation's connection to intrinsic motivation (Sects. 3.2.3 and 3.2.4), extrinsic motivation, and the region's social capital level are outlined (Sect. 4.3). The recommended implementation method (Sect. 3.3.2), measurement of companies' social responsibility degree (Sect. 3.3.4), and assessment of actors' satisfaction with the implementation (Sect. 3.3.5) are discussed. Furthermore, the comparison between declared social responsibility content and its actual implementation (Sect. 3.3.6) and the economics of implementing social responsibility (Sect. 3.3.7) are addressed. Section 3.3.8 offers a basic overview of knowledge and findings on social responsibility, along with related research topics, in the form of an annotated mind map.

3.3.1 Human Value System and Socially Responsible Behaviour

As illustrated by González-Rodríguez, Díaz-Fernández and Simonetti [52], human values influence perceptions of social responsibility, and the way consumers perceive social responsibility mediates the relationship between human values and the way entrepreneurs perceive social responsibility. They also state that significant differences can be found in consumers' and entrepreneurs' perceptions of social responsibility with respect to cultural background. This link between human values and socially responsible behaviour is part of the theory of planned behaviour model to investigate environmentally friendly behaviour, including willingness to pay for the quality of the environment (e.g. [38, 53]), also referred to as the environmental theory of planned behaviour model by some authors (e.g. [54]). Kim and Han [26] extended this model to a modified model of the theory of planned behaviour by adding value-behavioural and perceptual aspects of human behaviour (see Sect. 3.2.3).

3.3.2 Recommended Way of Implementing Social Responsibility

With the intention of increasing its effectiveness, corporate social responsibility should be introduced and implemented in a planned and strategic manner, as Grayson and Hodges [55], cited in Ma [56] state in their seven steps of initiation, gradual implementation and development:

1. Initiating business case/situation.
2. An assessment of the organisation's resources, including its social capital and the motivation of individual employees and management.
3. Identification and evaluation of motives/motivations for CSR commitment, including an analysis of the economic benefits of its implementation; see, for example the proposed Cost-Benefit Analysis (CBA) method for airlines ([57]).
4. Practical implementation of social responsibility.

5. Strategic approach to corporate social responsibility.
6. Stakeholder recognition and engagement.
7. Measurement, evaluation and reporting.

Prior to the introduction of social responsibility in an organisation, a situational analysis focusing on the organisation itself, its environment and competitors should be carried out to increase the effectiveness of its implementation, just as it is done in management and marketing before a major strategic decision (e.g. [58, 59]). This situational analysis should also include, in addition to the resources and motivations for introducing social responsibility in the organisation, the following:

- Motivation for implementing social responsibility, its results, the benefits and costs of its implementation for competitors and collaborating organisations, typically the organisation's suppliers.
- Demand it from potential customers, the organisation's environment and the public.
- An estimate of future social demand for social responsibility in the organisation, in the destination and at national and global levels.

According to Vlachos et al. [60], management of organisations should also monitor consumer trust, which significantly influences consumer participation in social responsibility activities.

In managing the relationship between consumers and organisations, the implementation of social responsibility should be based on research conducted by Becker-Olsen, Cudmore and Hill [61]. According to them, consumers' perceptions are negatively affected not only by poorly targeted social responsibility activities, but also by well-targeted social responsibility activities that are timed and can thus be perceived by consumers as advertising. On the other hand, according to this research, consumers react positively to well-targeted social responsibility activities that are proactively designed.

According to Argandoña [62]:15–16, the recommended approach to implementing social responsibility for tourism operators includes seven basic steps and principles (Box 3.2).

Box 3.2 Recommended Procedure for the Implementation of Social Responsibility for Tourism Entities. Source: [62]:15–16; Extended, Modified

1. Definition of the desired scope of social responsibility and readiness for its fulfilment even though the short-term benefits of its implementation may be minimal for the tourism entity.
2. Social responsibility requires a long-term vision, which, especially for smaller tourism enterprises, requires development.
3. Setting a corporate culture and rules for implementing social responsibility using training and competence enhancement of internal and external tourism actors (see, e.g. Chou, 2014; for hotels).

4. Preference for cooperation with other tourism actors, in particular, NGOs, local government, tourism experts and local people, rather than implementing social responsibility autonomously.
5. Draw on the experience of other tourism actors.
6. Not to strive for maximum precision and immediate implementation of all intended social responsibility activities.
7. Be realistic about the scope of implementing social responsibility and consider the costs, especially for small companies operating in very competitive markets.

3.3.3 Formalisation of Social Responsibility Within ESG

The concept of social responsibility has evolved similarly to the evolution seen in the transition from informal quality achievement and environmental impact reduction within companies and organisations to the formalisation of standards like ISO 14000+ and integrated management. The formulation of the ESG (Environmental, Social, Governance) concept describes the degree of application of social responsibility in organisations in a more structured and concrete manner. Since approximately 2006 ([63]), the number of publications on ESG has been gradually increasing, and a dynamic professional interest can be noted since 2015. The ESG concept encompasses the economic aspects of social responsibility for companies and organisations, focusing on three main categories of social responsibility realisation ([64, 65]), whilst the emphasis on social responsibility is expressed by two separate categories of social and administrative:

- **The environmental category** emphasises prevention with investments in material and energy efficiency and in the minimisation and recycling of waste, which brings economic benefits in the long term. The evaluated metrics within this category can be, for example the rate of emissions and waste production, the rate of waste recycling, impacts on ecosystems and biodiversity, energy consumption and the efficiency of its use, and the efficiency of their products in relation to the environment.
- **The social category** focused on good working conditions for employees and support of positively perceived activities in the company and society. The evaluated metrics within this category can be, for example the quality of employee training, the exclusion of child labour and discrimination in employment, safety and health protection in the workplace and support for the local community.
- **The governance category** emphasises compliance with ethical standards, transparency of own activities and effective management. The evaluated metrics within this category can be, for example the existence and quality of codes of conduct and business principles, transparency of business and the publication of reports, and methods for eliminating bribery and corruption.

According to Ionescu et al. [66], the governance category has the greatest influence on the value of a company in tourism, which is apparently related to the growing emphasis on the ethical aspects of companies' business.

3.3.4 Measuring CSR Results

In addition to using exact economic, environmental and socio-cultural data, the measurement of CSR performance can also be based on the subjective perception of these results by different tourism stakeholders. This perception can be evaluated both quantitatively and qualitatively, including the use of mind mapping.

In a quantitative assessment, CSR indices can be constructed as described, for example by Bonilla-Priego, Font and Pacheco-Olivares [67] and Martínez, Pérez and del Bosque [68]. Criteria for achieving a certain level of social responsibility can also be defined, as shown in general by, e.g. Martínez, Pérez and del Bosque [68] and also Matev and Assenova [69] to assess the level of social responsibility according to the ISO 26000 standard for organisations in Bulgarian tourism.

The corporate sustainability reporting index, constructed by Bonilla-Priego, Font and Pacheco-Olivares [67], is a very comprehensive indicator, comprising seven environmental categories (A1-A7), six social categories (B1-B6) and one economic category (C1). They further subdivided these categories by type into soft/hard and managerial/performance categories. This classification and purpose of each category is presented in the Box 3.3.

Box 3.3 Components of the Corporate Sustainability Reporting Index by Bonilla-Priego, Font and Pacheco-Olivares [67]

- Categories A1 and B1 focus on hard management measurements related to the structure and management systems of the organisation, put in place for the protection of the environment and social responsibilities.
- Categories A2 and B2 focus on credibility, obtained through external awards and certifications in EMS or social responsibility systems (mainly ISO 14001, EMAS, SA8000, ISO 26000 standards and guidelines).
- Category A3 includes performance indicators in relation to the environment.
- Category B3 refers to social responsibility towards society and the labour process and covers aspects of work including dignity of work, respect for human rights and product safety.
- Categories A4 and B4 are financial indicators for proactive environmental initiatives such as R&D expenditure, fines and employee training.
- Categories A5–A7, B5 and B6 are soft indicators and include in particular the organisation's environmental and social policy statements, strategies, codes of conduct and commitment to current and future legislation.
- Category C1 is dedicated to hard economic indicators, including general categories of economic performance and economic influence on destinations.

Bonilla-Priego, Font and Pacheco-Olivares [67] applied this index to the sea and ocean cruise market. They found that this segment behaves conservatively, applies social responsibility based on its own approaches and is not affected by the pressure of tourism actors. They found that, compared to other sectors, cruise companies scored very low on the economic and socio-cultural aspects of social responsibility.

De Grosbois [70] used a simpler approach to analyse the social responsibility results of the world's 150 largest hotel chains. He focused on five main areas (environmental goals, quality of jobs, diversity and availability, welfare of society and community, economic prosperity) and statistically evaluated the occurrence of suitable approaches in these areas (for more details, see Sect. 5.1).

3.3.5 Tourism Actors' Satisfaction with the Implementation of Social Responsibility

Tourism actors' satisfaction with the implementation of social responsibility comes, aside from their intrinsic motivation, mainly from the results of social responsibility in the organisation and in the destination, as well as from the way these results are communicated. The quality of the services provided also has an impact, as shown by Gao and Mattila [71]. The results of their comparative research on guest satisfaction with the services of environmentally friendly and conventional hotels showed that:

- If the service is provided in good quality, they appreciate it more in environmentally friendly hotels.
- However, if the service is not provided in good quality, the rating is the same for both types of hotels.
- Guests' relationship with environmentally friendly hotels is stronger and satisfaction with their services is higher if the hotels engage in public benefit activities than if their environmental activities are focused internally.

Chen [35] analysed the relationship between customers' choice of an environmentally friendly hotel and its amenities. She concluded that customers are most likely to be influenced by the level of implementation of recycling programs, offering locally grown and produced food, environmentally friendly cleaning methods, energy efficient lighting, water conservation and the use of recycled materials. These findings were summarised by Millar and Baloglu [72] by finding that guests who preferred an environmentally friendly accommodation (regardless of whether they were business or leisure travellers) considered third-party environmental certification as the most important reason for choosing accommodation. That is, out of a range of environmental attitudes and activities including recycling policy, shampoo facilities, lighting control, energy-efficient light bulbs, frequency of changing towels and linen, and third-party environmental certification.

The importance of the way in which social responsibility is communicated was examined by Dawkins [73]. As she states, the successful implementation of social responsibility is as important as the discussion of it with a variety of stakeholders. According to her, it is necessary to develop a communication strategy that is primarily concerned with the key components of social responsibility. This communication should, in the way it is implemented, be appropriately targeted at key stakeholders, taking into account their expectations, and should not forget the organisation's employees, who are crucial to the successful implementation of social responsibility.

3.3.6 Declared Content of Social Responsibility and Reality

Many studies (e.g. [74, 75, 76]) show that the degree and manner of actual implementation of social responsibility is often in considerable discrepancy with the declared one. This applies not only to tourism, where reasons for this discrepancy include, in particular, high competitive pressure, assumed economic costs (see Sect. 3.3.7) and insufficient government support ([75]).

Font et al. [74] in their research on hotel chains concluded that the level of social responsibility in hotel chains is still low. They specify that in the environmental area, it is only motivated by savings, the operational process only concerns compliance with local legislation, socio-economic policies are only internally oriented and do not consider the consequences at the destination, and customers are insufficiently involved in the social responsibility process. They also found a difference between large and smaller hotel chains. According to them, large hotel chains (most of which declare social responsibility; [77]) have adopted a more holistic view of social responsibility. They add that these large hotel chains show a large gap between the declared content of social responsibility and its actual implementation. They stress that smaller hotel chains tend to favour only environmental management but fulfil what they have promised.

Juvan and Dolnicar [76] found an interesting motivational–behavioural paradox amongst conservationists who are actively engaged in nature conservation in their home country. This paradox is that whilst on vacation, their behaviour has negative consequences for nature, which they themselves perceive negatively and also justify to other people. Young, Higham and Reis [78] reach similar conclusions regarding air passengers' psychological discomfort. As Juvan and Dolnicar [76] state, clarifying this paradox for conservationists could contribute to understanding the socially irresponsible behaviour of other tourism participants.

3.3.7 Economic Perspectives on Social Responsibility

The implementation of social responsibility also entails investments and other costs of the organisation for some activities. At the same time, these can lead to a significant increase in the quality of the provided product, including the moral and ethical level of its evaluation by the product provider and especially by its end user (see also Sects. 3.2.2 and 3.2.3). Various studies (e.g. Kang et al. [30] and Parsa et al. [31]) have investigated whether and to what extent customers are willing to pay more for improving the "moral" quality of a product related to social responsibility (see also Sect. 3.2.3). Similarly, this question of "willingness to pay" is asked in connection with the willingness to share in the costs of maintaining or improving the quality of the environment, which is one of the methods of environmental economics. This method is generally based on the finding of how much a resident (or other people) is willing to pay for preserving the quality of natural assets, and thus leads to their valuation. Parsa et al. [31] investigated the willingness to pay a higher price for more socially responsible services. They combined a questionnaire survey and face-to-face interviews with clients directly in restaurants. They found that clients are generally willing to pay only a relatively small price increase for the implementation of social responsibility activities by restaurants. However, their findings also indicate that clients who were more interested in social responsibility and/or directly involved in social responsibility activities were willing to pay more for socially responsible services. This indicates that these clients consider social responsibility as an added value to the offered service. Kang et al. [30] came to similar conclusions for US hotels and their environmentally friendly activities—more environmentally conscious guests are willing to pay more for environmentally friendly hotel activities. In addition, according to this study, guests of high-level and mid-range hotels are willing to pay more compared to guests of low-category hotels.

Well-conceived social responsibility practices, especially cleaner production methods, environmental management, promotion of recycling (see the potential benefits of recycling for hotels in Singh, Cranage, and Lee [79]) and green management can, in turn, bring economic benefits to organisations. This is confirmed by Kang et al. [80] for family-run hotels; the benefits of social responsibility for family-run hotels were agreed upon by all interviewees—hotel employees, hotel managers and hotel guests. This is in line with the general finding that family businesses show higher potential for social responsibility due to their connection with the local community (Niehm, Swinney, and Miller [51]).

In recent years, family businesses with features of small enterprises have been investigated in relation to the differences in management and economic performance, their role in rural areas and the specifics of social responsibility. As stated by Singal [81], it is a well-known finding (see, e.g. [51]) that family businesses have better economic performance and higher social responsibility compared to non-family businesses of comparable size and in the same business sector. Singal [81] concludes from research in the tourism and hospitality industry that:

- Family businesses in tourism and hospitality have better economic performance.
- Tourism and hospitality entrepreneurs do not invest so much in social responsibility (contrary to commonly accepted conclusions from other business

sectors), but rather invest in mitigating concerns about the negative impacts of non-compliance with social responsibility, such as loss of reputation.
- If these businesses invest prudently in social responsibility, they will reap significant economic benefits in the future.

3.3.8 The Basic Context and Knowledge of Social Responsibility and Related Research

As stated by Popa [82], research on social responsibility has received a great deal of attention in the last quarter of a century in relation to the management and marketing of organisations and business ethics. A comprehensive mind map (Fig. 3.8) brings a basic context and knowledge regarding social responsibility, topics and other aspects of its research and application, with an emphasis on the structured nature of reflections on social responsibility. Apart from our own research and literature review ([43]:782) this mind map and Box 3.4 are based on the findings of many other authors: [31, 73, 74, 83, 84, 85, 86, 87, 88, 89, 90, 91, 92]. This mind map tries to respond to the key questions related to social responsibility (Box 3.4).

Box 3.4 Key Social Responsibility Questions
- What is the essence of social responsibility?
- What formal or informal approaches can be used to implement social responsibility?
- How does social responsibility affect different types of organisations, including the employees who implement it?
- In relation to a particular organisation/destination, what are the implications of declared and actually implemented social responsibility?
- What conditions are involved in the implementation of social responsibility, or what influences its effectiveness?
- What is the motivation for its implementation?
- How should the communication about its intentions and the way of implementation take place?
- What are the barriers, risks and complications connected to its implementation and how to avoid them?
- How does implementation and effectiveness vary in different contexts?
- How and to what extent will customers support social responsibility?

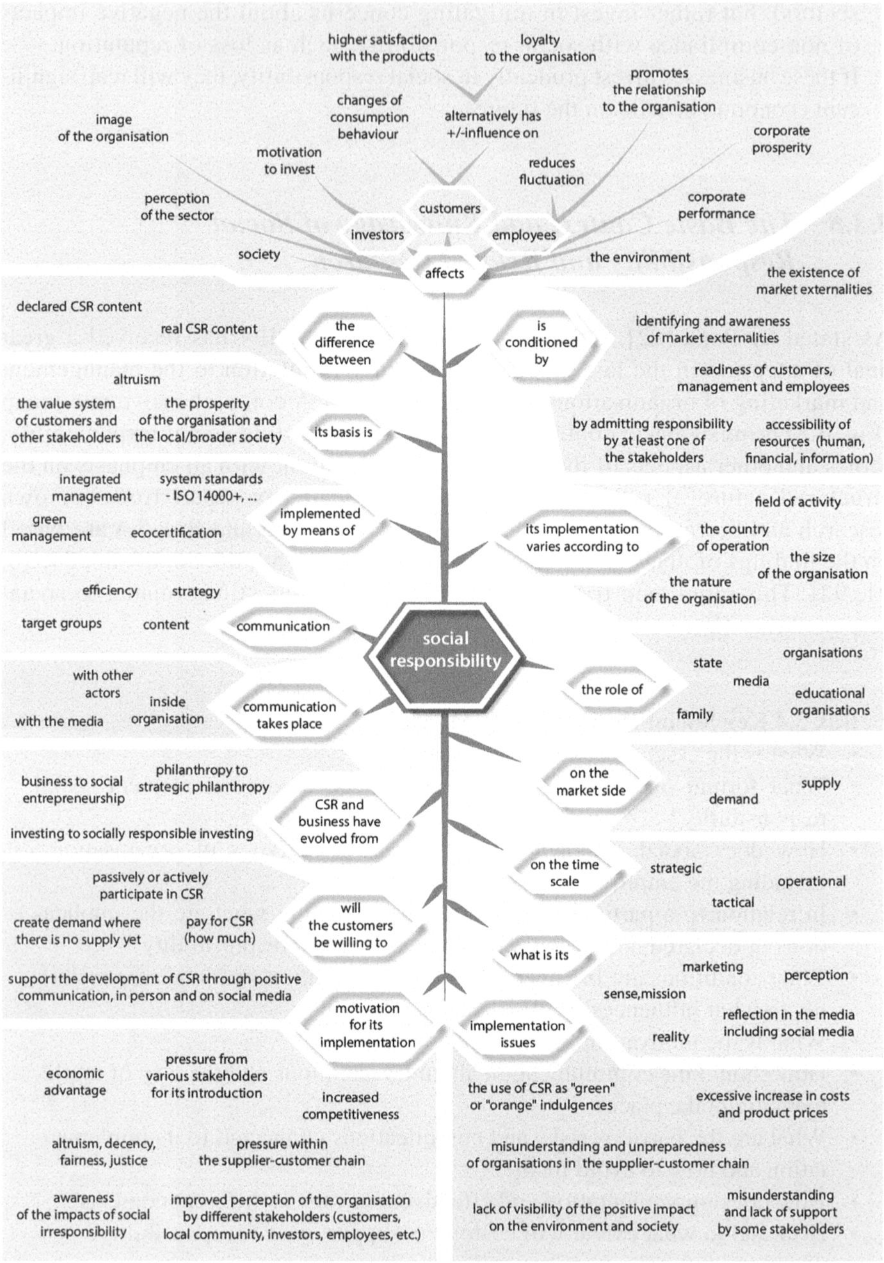

Fig. 3.8 A mind map of the basic context and knowledge regarding social responsibility and related research. Source: [43]:782 and the authors mentioned in the text commenting on the picture

3.4 The Impact of Social Responsibility on Different Tourism Actors

The declaration and implementation of social responsibility activities have different effects on tourism stakeholders and destinations under different conditions and in conjunction with different tourism actors. As research on these effects has shown, similar to market externalities, their consequences are often difficult to predict and can be both positive and negative, or intended and unintended consequences. These can be observed in changes in motivation, expectations, emotional evaluation, decision-making and in changes in the behaviour of tourism actors. The degree and nature of changes induced by the implementation of social responsibility are also determined by the nature of tourism actors, including their individual and group differences, current emotional attitudes and cognitive levels. The following chapters analyse in more detail the impact of social responsibility on employees of organisations and clients of tourism services providers.

3.4.1 The Impact of an Organisation's Social Responsibility on its Employees

Employees significantly influence the successful implementation of social responsibility and at the same time social responsibility affects employees. Research on this relationship can contribute to understanding the quality and extent of this influence, to understanding the most important components of social responsibility in terms of its impact on employees, and to finding the employee characteristics that influence this relationship. In this context, Wells et al. [93] highlight the importance of combining qualitative and quantitative research. The key aspects of this relationship are presented in the mind map in Fig. 3.9.

The impact of social responsibility on an organisation's employees is much less studied than the impact of social responsibility on potential clients. In a superficial assessment by corporate management and probably in the thinking of academics as well, quantitative research seems more relevant, especially in terms of the impact of implementing social responsibility on client decision-making, and is also easier to conduct. Chan and Hawkins [96] analysed the impact of EMS implementation on employees and management of hotels in China. They concluded that EMS implementation has both positive and negative effects on employees and that a top-down approach is a more appropriate way to implement EMS in China, also due to the cultural environment. According to them, positive effects with great motivational potential for further support of socially responsible activities include better, safer and healthier working conditions, a feeling of positive impact on society, pride in obtaining eco-certification, higher job satisfaction and increased value of employees on the labour market, including training, education and work experience in an environment with an established EMS. In their

Fig. 3.9 A mind map of the impact of an organisation's social responsibility on its employees. Source: Inspired by [94, 95]

research, employees perceived EMS negatively if it was introduced only to save money, improve the company image and attract environmentally conscious guests. This superficial approach is often referred to as "greenwashing". Thus, in the case described by them, the basis of EMS was not the adequate social responsibility approach of the hotel management and its employees.

In relation to the implementation of green management in hotels, Chan et al. [97] found a positive correlation between three attributes of employees (environmental knowledge, environmental awareness and environmental concern) with their environmental behaviour as well as with their support for the implementation of green management in hotels. At the same time, according to Wells et al. [93], it is important to place much emphasis on adequate employee motivation, perception of the potential for change and appropriate levels of training which will ensure that employees' environmental preferences are satisfied and consistent with their environmentally friendly behaviour.

Farooq, Farooq and Jasimuddin [98] came to a fundamental finding about the impact of social responsibility on employees. They found that different components of social responsibility affect employees differently according to their character traits and social behaviour. This could be explained by a model similar to the theory of planned behaviour (Sect. 3.2.3). Their important finding shows that individualistically oriented employees are more influenced by activities aimed at the local community, whilst collectivistically oriented employees are more influenced by activities aimed at the organisation's employees. According to Singhapakdi et al. [99], the harmony between an organisation's social responsibility and content preferred by the organisation's employees is also critical to the effectiveness of social responsibility, otherwise, the employees feel dissatisfied with their own work in various ways.

The perception of social responsibility of hotels also affects potential employees. According to Yilmazdogan, Secilmis, and Cicek [100], this perception positively affects tourism students' decision to work in the field. According to them, this resolution also correlated positively with partial components of social responsibility, specifically with ethical, legal and economic responsibility as well as philanthropy.

3.4.2 Impact of Social Responsibility on (Potential) Clients

The influence of social responsibility on both potential and actual clients varies according to the different segments as it reflexes their value system, needs and travel experiences (Fig. 3.10). Selected relevant effects and factors of these impacts are analysed in the following chapters, namely:

- Willingness to pay for an increase in product quality by introducing social responsibility—Sect. 3.2.3.
- The impact of social responsibility on both potential and actual customers later in this chapter—environmental component of social responsibility.
- The relationship between the individual system of values and social responsibility—Sect. 3.3.1.
- The impact of social responsibility on the satisfaction of actors (including clients) with the implementation of social responsibility—Sect. 3.3.5.

By analysing statistical data on social responsibility and customer satisfaction of hotels and restaurants in the USA, Lee and Heo [102] verified that positive social responsibility activities in hotels positively affect guest satisfaction and the value of the relevant company. It is important that the content of social responsibility intended and implemented by management corresponds to the needs and preferences of customers. As Nasution and Mavondo [103] discovered, hotel managers' ideas about customers' values can differ from the values perceived by the customers themselves.

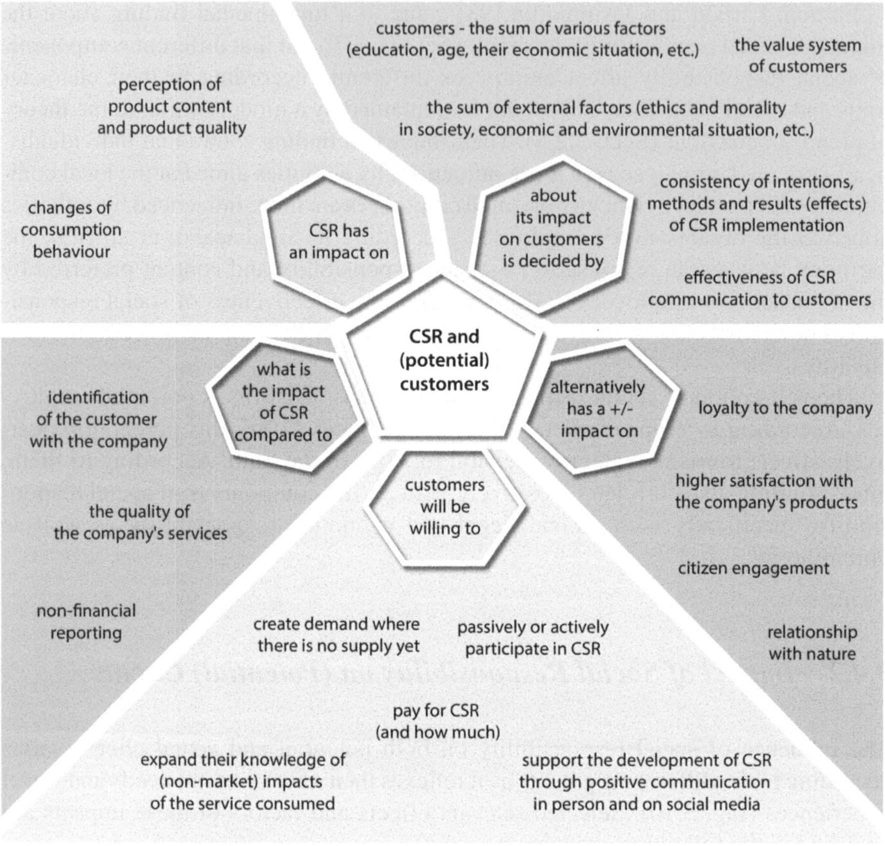

Fig. 3.10 Mind map of the influence of the implementation of social responsibility on (potential) clients. Source: Inspired by [101]

3.5 Selected Approaches, Methods and Concepts of Social Responsibility

In practice, social responsibility is based on the types of approaches, methods and concepts shown for illustration in the diagram in Fig. 3.11, alternatively or in their various combination and synergies. For these approaches, typical methods of their implementation are also briefly presented. At the same time, the method of implementation may, depending on its complexity, include several types of approaches, methods and concepts.

1 **Information Analytic Approach**

The approach of information analytics includes non-financial reporting of companies, tourism sustainability monitoring and environmental accounting. It is the basis for communication amongst tourism actors and for decision-making in

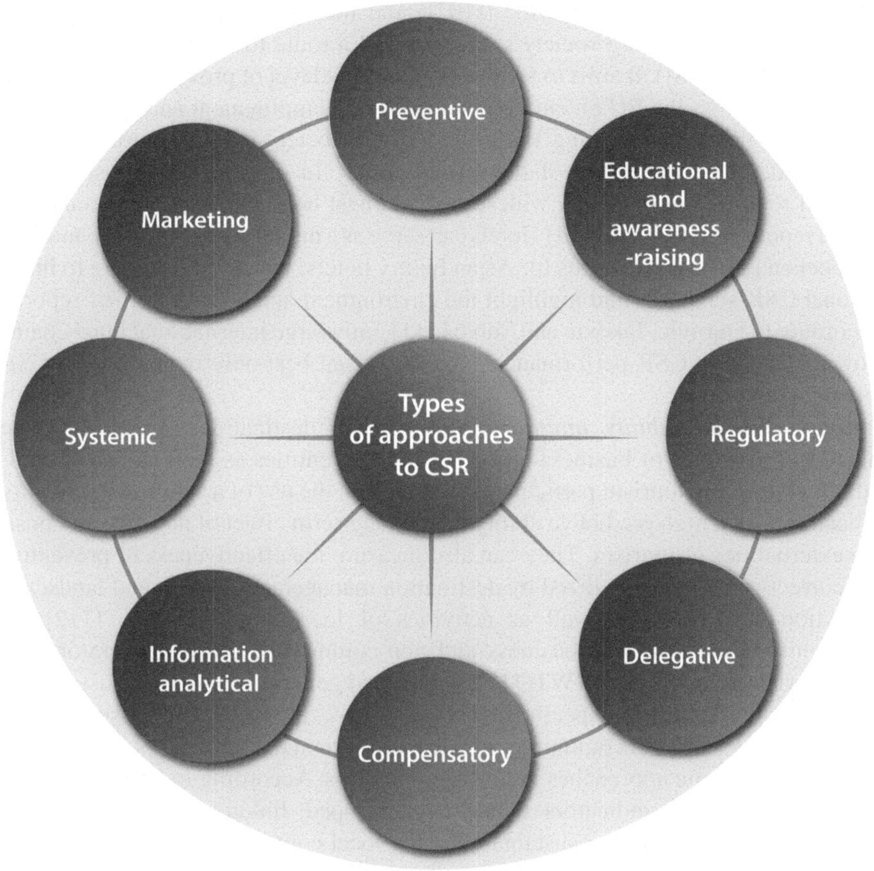

Fig. 3.11 Schematic representation of the types of approaches to the implementation of social responsibility

management, especially tourism service providers and destination management. It also includes the choice of tourism services and destinations by tourism participants. The growing importance of CSR related to the integration of social responsibility into corporate strategy is highlighted, for example by Fernandez-Feijoo, Romero and Ruiz [104]. They consider the integration of CSR into the basis of business as important for the interaction with stakeholders and perceive sustainability reports as key communication tools for companies.

The non-financial reporting of companies presents the intended content of corporate social responsibility and, in particular, the achieved results, which provides an opportunity for control by tourism actors. Since 2014, the EU Non-Financial Reporting Directive ([105]) has been in force for large companies ([106]). The Directive follows the "European Parliament resolution of 6 February 2013 on corporate social responsibility: accountable, transparent and responsible business behaviour and sustainable growth" ([107]) and the document "Corporate Social Responsibility: promoting society's interests and a route to sustainable and inclusive recovery" ([108]). It aims to achieve a consistent level of provision of corporate information across the EU on environmental, social, employment and human rights issues. Non-financial reporting is thus a tool for society-wide monitoring of the level of social responsibility of reporting entities. In the tourism industry, non-financial reporting is relatively widespread amongst hotel chains, and the environmental reporting activities of NH Hotels can serve as a model ([109]). Cherapanukorn and Focken [110] illustrate this for Asian luxury hotels, which often adhere to international CSR standards and highlight the environment and society in their reports. According to Đurović, Jaković and Tubić [111], only large international hotel chains provide details on CSR performance; small hotels at best only mention their CSR vision and mission.

Tourism sustainability monitoring in tourism destinations, monitoring the impacts of activities of business and other tourism entities as well as that of individuals (especially tourism participants) is based on the use of a set of sustainability indicators with a high predictive ability on the long-term effect of negative and positive externalities of tourism. They can also measure the effectiveness of preventive and corrective activities realized by destination management, nature and landscape protection authorities, as well as activities of local residents (e.g. [112] for community-based tourism), business and non-commercial tourism operators and other tourism actors. The UNWTO [113] designed general and specific sustainability indicators, destination-specific indicators ([114, 115, 116]) for assessing the environmental and local community impacts of tourism and indicators of the effectiveness of systemic approaches in tourism operators. According to Franzoni [117], more comprehensive indicators should be developed, linking, in addition to the three dimensions of tourism sustainability, the local community, the tourism destination and the local tourism organisation. For tourism operators, according to Hsiao et al. [118], the ecological footprint approach (e.g. [119]) is being developed for the application of EMS in hotels, with an assessment of the impact of tourism on resources, or related to environmental impacts (e.g. carbon footprint—[120, 121, 122]). Wang et al. [123] derived standards for green management of restaurants based on the green supply chain management (GrSCM) of restaurants and their supply chains. Using the Delphi method, they arrived at a set of 81 sustainability indicators. Gallardo-Vázquez and Sanchez-Hernandez [124] also introduced CSR indicators for companies.

Environmental accounting, based on the concept of ecosystem services ([104, 125, 126]), complements "classical" accounting by adding free (non-scarce) goods (e.g. [127]). Its utilisation in tourism sustainability is discussed in general terms by, e.g. Garrod and Fyall [128]. A specific application is provided by Raybould et al. [129] in case studies appraising the value of Australian beaches and the willingness of visitors and residents to pay for their maintenance. Gössling et al. [130] showed the economisation of free goods in evaluating the environmental economic efficiency of selected destinations, and Law et al. [131] in modelling the transition to a green economy in the destination of Bali, Indonesia.

2 Systemic Approach

The basis of the systemic approach is to systematically, continuously and thoughtfully manage the processes, system elements and resources of the destination or organisation. In organisations, the systemic approach implies the implementation of consistent changes to the management of the whole organisation. Alternatively, these changes can be based on the application of international ISO standards, EU directives or on the approaches and methods of TQM (Total Quality Management) and TQEM (Total Quality Environmental Management).

The implementation of system solutions using international standards and EU directives focuses on environmental management (ISO 14000+, EMAS, see Sect. 4.8), integrated management linking a Quality Management System (QMS), an Environmental Management System (EMS) and an occupational Health and Safety Management System (HSMS) and corporate social responsibility with criteria for production process management (SA8000 standards—see Sect. 4.6, ISO 26000—see Sect. 4.5). Integrated management, when applied according to ISO standards, progressively combines the use of ISO 9000+, ISO 14000+ and ISO 45001:2018 (former OHSAS 18001, Occupational Health and Safety Assessment Series). An important part of the ISO systemic approach is prevention.

The system solution can also be ***based on TQM methods*** (e.g. [132] for hotels) with an emphasis on quality focused on environmental management (TQEM).

3 Compensatory Approach

In the compensatory approach, the negative externality is compensated at the point of origin or at an appropriate location by an activity that eliminates or reduces the damage caused or by a socially beneficial activity (discussion for tourism, e.g. [62]), Fig. 3.12.

At the point of origin, the negative externality can be counterbalanced by activities that improve the quality of the environment and the quality of life of the local community. This can in turn become a positive externality for tourism participants or other tourism actors, as it increases the quality of the destination or its attraction or infrastructure. For example, an airport can counterbalance the noise pollution caused by aircraft by funding the planting of green belts in nearby residential areas and by reimbursing the local residents for thermal and acoustic insulation windows (e.g. [133]).

Compensation benefiting the entire society is typical for cases where the negative externality is non-local, e.g. aircraft emissions. As a result, offset programmes organised and funded by airlines, and co-financed by passengers (offset programmes Lu and Shon [134] and Chen [83], critical evaluation of their effectiveness by Higham et al. [32]) are carried out in different parts of the world, especially in poor developing countries. For example, these may include rescue programmes for endangered species (e.g. mountain gorillas, rhinos, and elephants), replanting forests in Africa, support for renewable energy (e.g. [135]) and educational programmes on sustainable forest management.

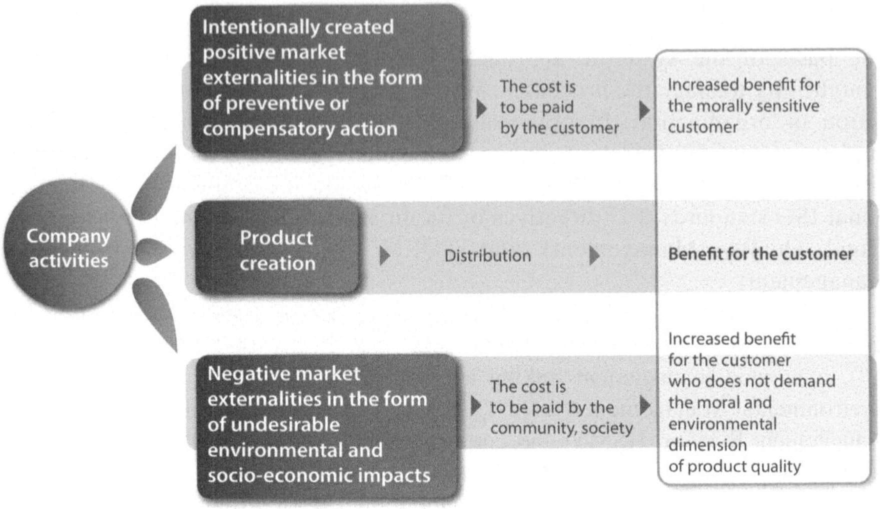

Fig. 3.12 Company activities in the context of market externalities and social responsibility

4 **Delegative (Client) Approach**

The delegative or client approach is based on the fact that the institution fully or partially delegates responsibility, including environmental and social impacts (e.g. [40]).

In the case of **full delegation**, the environmental and social costs are calculated in the price of the product.

In the case of **partial delegation**, the client contributes voluntarily to the prevention or compensation of environmental and social costs. This includes various forms of direct involvement of tourism participants in social responsibility activities. In the tourism industry, these include offset programmes and the involvement of clients in the environmentally friendly activities of hotels, for example their green management and waste management programmes. Singh, Cranage, and Lee [79] conducted a cost-benefit analysis for waste management, quantified the benefits of recycling in reducing greenhouse gas emissions, and recommended the involvement of hotel guests in recycling programmes.

5 **Preventive Approach**

With a preventive approach, the company strives to eliminate the adverse effects of its activities on society and the environment throughout the product life cycle. This is done by applying various practices and solutions, which include fair trade principles ([86, 136, 137, 138]), the implementation of EMS, TQEM and cooperation with suitable suppliers, e.g. Wyngaard, de Lange [139] confirm the interest of hotel management to cooperate with eco-farms. The company tries to eliminate its negative impacts even if it means increasing costs, i.e. investments in human resources and innovation. A responsible preventive approach is essential for the long-term sustainability of tourism. It includes strategic planning, long-term reduction in the consumption of non-renewable resources (e.g. [135]), management approaches to increasing carrying capacity of tourism destinations or its special localities, improving the quality of tourism infrastructure, training and education of all tourism actors, long-term efforts to change the mindset and behaviour of tourism actors, and well-defined roles of tourism actors in the application of preventive measures.

6 **Educative-Awareness Approach**

The educative-awareness approach focuses on the application, presentation and clarification of principles, which make clients engage in sustainable behaviour. Geoparks (Sect. 6.1), facilitating educational opportunities, can serve as a comprehensive example. Their basic premise is holism, linking the interpretation of geological heritage, flora and fauna, cultural heritage with the lifestyle and customs of local people (see [140, 141]).

7 **Regulatory Approach**

The basis of the regulatory approach is the establishment of geographically, quantitatively or qualitatively conditioned rules for the use of the territory, its resources and processes. It is primarily concerned with the spacetime distribution and regulation of the intensity and/or form of tourism, particularly in naturally and socio-culturally vulnerable areas, including restricting or prohibiting access to certain sites and trails. It is also necessary to regulate the operations of tourism operators, including, on the input side, the limitation of materials used (e.g. coming from endangered species) and the regulation of the use of human capital.[6] The same pays for the output side, e.g. the limitation of emissions of hazardous substances into the environment or the regulation of landscape interventions. These regulations are imposed:

- **Strictly**—Usually legislation for protected areas, emission limits (e.g. [62]), fixed "administratively declared" carrying capacity (e.g. 50 visitors per day to visit the Roraima Table Mountain, Venezuela/Brazil/Guyana).
- **Using the assessment of** (even unintended) **impacts**—in tourism, the concept of limits of acceptable change and a number of visitor management approaches based on maintaining the quality of visitor experience (detailed in [18, 45, 142, 143]).
- **Dynamically with the knowledge of natural (and social) processes**[7]—dynamically applied the concept of carrying capacity ([45, 48]).
- **Administratively and economically using taxes**—typically spa and recreation fees (e.g. [40, 144]).
- **Using market mechanisms**—price regulation based on the price elasticity of demand, e.g. visitor card.

---

[6] For example, inadequately remunerated, unsafe, undignified or otherwise disadvantageous working conditions for certain population groups, especially local and Indigenous peoples.

[7] For example, bat hibernation cycle in caves, nesting birds.

The effects of regulation should be anticipated and modelled in advance, as demonstrated, for example by Tol [144] for modelling changes in the tourism industry in relation to the possible introduction of a tax on aviation fuel and the consequent reduction in CO_2 emissions. In relation to tourism, it is important to ***set and enforce certain rules to eliminate negative externalities***. A typical example is the designation of large- and small-scale protected areas and the protection and rules for the use of tangible cultural and technical heritage. In the national economy, ***the use of market-based regulatory instruments*** aimed at economic efficiency of the market is growing. For example, emission allowances are being introduced to reduce CO_2 emissions in the energy sector, and their introduction in transport is also planned.

The ***concept of carrying capacity*** is increasingly being used for regulation of tourism intensity in a given area. It is used predominantly in its static form, most often in the form of the fixed limits of the territory use based on expert opinion on self-regenerative power of ecosystems and social systems impacted by tourism. In the future, it will also be used in its dynamic form ([45]). The use of carrying capacity is based on the commitment to maintaining the desirable state of the territory, the living conditions of the local population and the experience of tourism participants. In this way, the application of this concept is an expression of the responsibility of the current generation to ensure the same conditions for future generations to satisfy their needs concerning quality of life and travel experience. And this is exactly the essence of tourism sustainability.

However, as very little is currently known about the complex interplay between the activities of the various tourism actors in the area (visitors, transport operators and tourism service providers), it is extremely difficult to use the concept of carrying capacity. Moreover, these actors are influenced by the characteristics of the territory, the weather, the history of their activities and the resulting consequences, which further hinders the practical application of the concept of carrying capacity. Therefore, in practice, it is more convenient to combine the decisions of destination management and tourism actors, which focus on the necessary regulation of tourism activities, with the concept of limits of acceptable changes, which is based on the evaluation of appropriate indicators of tourism sustainability. This approach is used in some visitor management models such as VIM, PAVIM, ROS and VERP ([18, 45, 142]).

8 Marketing Approach

The certification of environmental or social quality, linked to brand marketing, provides standardisation of social responsibility and often also of other aspects, such as the quality of services provided, in accordance with the certification scheme. According to Buckley [145], brand marketing should be associated with higher effectiveness of certified brands and certification schemes for local products, environmental quality of beaches, marinas, boats, hotels and hotel chains along with continuous improvement of environmental quality for EMAS and ISO 14000+ certifications. The certification of environmental or social quality is also important in terms of marketing—tourism participants increasingly base their decisions about purchasing a service, choosing a destination or a place in a destination on this declared and certified minimum social responsibility.

Some tourism actors' activities generate benefits as expressed in terms of market price, but also harm or costs to other tourism actors who did not cause them. For example, the increased emissions of pollutants as well as noise and light pollution caused by tourism have a negative impact on the local population in terms of increased health care costs and reduced income as a consequence of their higher vulnerability to illness. This is generally referred to as a negative externality. If such externalities are not internalised by including the cost of the negative impact (e.g. damage to the local environment) in the price of the tourism product, the cost is borne by local residents who did not cause it. These costs can be passed on to the visitor by increasing the price calculation of the tourism product. Either a precautionary approach to social responsibility or a compensatory approach can be applied. These approaches can include the purchase of low-emission and low-energy appliances and equipment, investment in green management in general, or investment in the revitalisation of the damaged environment. Once the market externality is included in the price, either by including the costs of eliminating or reducing negative impacts or by compensating for them, the responsibility for these costs is acknowledged and shouldered by the creator and consumer of the tourism product. However, externalities can also take a positive form. For example, local residents benefit from the improved quality of the environment, which the operator of an ecotourism product has achieved in order to increase the desirability of its offer. Nonetheless, the costs of this investment are borne by the operator and its clients.

3.6 The Role of Social Responsibility in Tourism Sustainability

The success of tourism is based on the quantity and quality of the attractions located in a destination. These attractions have been created independently of tourism and represent a more or less significant heritage for society, habitats and ecosystems, providing a life-sustaining environment for living organisms, including humans. They are therefore referred to as primary destination resources. In order to activate the potential of tourism, which these attractions represent, the prerequisites to tourism development are created in the form of tourism infrastructure (secondary destination resources) and the selective prerequisites of tourism (legal, political, security environment etc.) also play an important role. The building and operation of tourism infrastructure have an impact on the quality of life of local residents, on cultural and technical heritage and on the environment. As a rule, the costs and benefits associated with these effects are not calculated in the price of the tourism product, they are market externalities (see Sect. 3.2.5).

If tourism in a destination is carried out in a sustainable manner, the positive effects outweigh the negative ones. This is due to the fact that the tourism actors in the destination behave responsibly, i.e. by showing a high level of respect for the life and environment of the local community. For this reason, they prefer approaches involving precautionary and/or compensatory measures for the externalities of tourism market and a sufficient level of social responsibility. Thus, in an optimal case, the destination prevents and compensates for the negative effects of tourism. Avoidable adverse effects that would lead to negative externalities of the tourism market are eliminated. At the same time, unavoidable negative effects are sufficiently offset by positive market externalities. Acceptance of their share of responsibility by individual tourism actors in the destination is stimulated and coordinated by appropriate, planned and conceptual development and use of destination resources as well as by the participation of tourism actors, especially by the local community, in destination management.

A major shift towards tourism sustainability would be manifested by a higher degree of declared social responsibility[8] and especially social responsibility implemented by tourism businesses, destination managements, visitors and other tourism actors, as shown by the research on the state of corporate responsibility in tourism management ([146]). These authors also note that research on social responsibility in tourism is still underdeveloped. In general, according to Egri and Ralston [147], research on social responsibility of organisations, according to an analysis of published articles, focuses on four basic themes (Fig. 3.13): corporate social responsibility (as part of social responsibility), corporate responsibility towards the environment, corporate ethics and corporate management.

Social responsibility is realised by using various approaches of environmental management[9] and cleaner production approaches ([148, 149]), environmental certification ([150]), environmental accounting, use of ecosystem services assessment ([126, 151, 152]) and local product labelling.

It is important to consider the manner, extent and effects of vertical interdependence of tourism sustainability at different geographical levels, as argued, for example by Pásková [14], as well as the role of external factors and external tourism actors (geographically external—outside the destination, e.g. [153, 154]). In line with many discussions on the extent to which the local community can influence the development of tourism in the context of decision-making by external actors including tourism infrastructure investors, politicians and tour operators, Tosun [155] concludes that the sustainability of tourism at the local level depends on appropriate national policies and on the activities of international bodies. International tour operators and international hotel chains are typical international players with significant influence in many destinations. For tour operators, Xin and Chan [156] analysed responsible tourism indicators, Font and Lynes [157] stress the application of social responsibility in their entire package. Budeanu [153] and Sigala [158] highlighted the responsibility of tour operators to promote tourism sustainability, and Lovelock [159] specifically stressed their ethical responsibility in planning and selling trips to selected destinations.

[8]Visions, strategies, policies, intentions, charters, codes of conduct and codes of ethics.

[9]e.g. according to ISO 14 001, EMAS, green management.

Fig. 3.13 Mental map of the desired structure and implementation of the social responsibility realised by a tourism entity

Concerning international hotel chains, Henderson [160] points to the important role of the growth in social responsibility of hotel chains as a factor in the recovery of destinations in Thailand after the 2004 tsunami.

References

1. Font, X., & Lynes, J. (2018). Corporate social responsibility in tourism and hospitality. *Journal of Sustainable Tourism, 26*(7), 1027–1042. https://doi.org/10.1080/09669582.2018.1488856
2. Marriott. (2022). Marriott business conduct guide, p. 46, [online]. Retrieved August 04, 2022, from http://www.marriott.com/Multimedia/PDF/CorporateResponsibility/Marriott_Business_Conduct_Guide_English.pdf
3. Marsden, K., et al. (2014). Integration of biodiversity into CSR processes in tourism, p. 44, [online]. Retrieved January 14, 2024, from https://adelphi.de/system/files/mediathek/bilder/csrbiodiv-baseline_study-barrierfrei-en_final_0.pdf
4. ISO 13810:2022: Tourism and related services. Visits to industrial, natural, cultural and historical sites. Requirements and recommendations.
5. Lee, S., & Heo, C. Y. (2009). Corporate social responsibility and customer satisfaction among US publicly traded hotels and restaurants. *International Journal of Hospitality Management, 28*(4), 635–637. https://doi.org/10.1016/j.ijhm.2009.02.007

6. Arnaudov, K., & Koseska, E. (2012). Business ethics and social responsibility in tourist organizations in terms of changing environment. *Procedia – Social and Behavioral Sciences, 44*, 387–397. https://doi.org/10.1016/j.sbspro.2012.05.043
7. Cohen, E. (1972). Towards a sociology of international tourism. *Social Research, 39*(1), 164–182. https://www.jstor.org/stable/40970087
8. Weaver, D. B. (2006). *Sustainable tourism: Theory and practice* (p. 240). Elsevier. ISBN 978-0-7506-6438-7.
9. Kim, K., Uysal, M., & Sirgy, M. J. (2013). How does tourism in a community impact the quality of life of community residents? *Tourism Management, 36*, 527–540. https://doi.org/10.1016/j.tourman.2012.09.005
10. WTAY. (2014). WTAY code of conduct, [online]. Retrieved August 22, 2022, from http://wtay.com/resources/code-of-conduct/
11. Tourism Sustainability Group. (2007). Action for more Sustainable European Tourism. Report of the Tourism Sustainability Group. http://www.ernestproject.eu/coalap/ernest/other_docs/10072009160632789_tsg_final_report_en.pdf
12. Černec, M. (2014). Corporate social responsibility practices in international hotel chains: A web content analysis. Dissertation, p. 126, [online]. Retrieved February 09, 2016, from http://www.cek.ef.uni-lj.si/magister/Cernec1527-B.pdf
13. COMMITTEE. (2002). Public participation in protected area management best practice. Parks and Wildlife Commission of the Northern Territory, [online]. Retrieved August 9, 2022, from https://www.environment.gov.au/system/files/resources/45209a20-1ab1-40d1-aa26-cf4838948782/files/public-participation.doc
14. Park, S.-Y., & Lee, S. (2009). Financial rewards for social responsibility. A mixed picture for restaurant companies. *Cornell Hospitality Quarterly, 50*(2), 168–179. https://doi.org/10.1177/1938965509331814
15. Buhalis, D., Papathanassis, A., & Vafeidou, M. (2022). Smart cruising: Smart technology applications and their diffusion in cruise tourism. *Journal of Hospitality and Tourism Technology, 13*(4), 626–649. https://doi.org/10.1108/JHTT-05-2021-0155
16. Zambrano, A. M. A., Broadbent, E. N., & Durham, W. H. (2010). Social and environmental effects of ecotourism in the Osa Peninsula of Costa Rica: the Lapa Rios case. *Journal of Ecotourism, 9*(1), 62–83. https://doi.org/10.1080/14724040902953076
17. Horng, J.-S., Hsu, H., & Tsai, C. Y. (2018). An assessment model of corporate social responsibility practice in the tourism industry. *Journal of Sustainable Tourism, 26*(7), 1085–1104. https://doi.org/10.1080/09669582.2017.1388384
18. Zelenka, J. (2015). *Cestovní ruch. Marketing [Tourism. Marketing]* (p. 247). Gaudeamus Hradec Králové. ISBN 978-80-7435-543-1.
19. Pásková, M., & Zelenka, J. (2016). Social responsibility role in tourism sustainability. In P. Jedlička (Ed.), *Proceedings of Hradec economic days, February 2nd and 3rd*, ISSN 2464-6059 (pp. 777–785).
20. Subway. (2022). Subway social responsibility, [online]. Retrieved August 10, 2022, from https://www.subway.com/en-US/AboutUs/SocialResponsibility, http://www.subway.com/subwayroot/about_us/Social_Responsibility/
21. Ramdas, M., & Mohamed, B. (2014). Impacts of tourism on environmental attributes, environmental literacy and willingness to pay: A conceptual and theoretical review. *Procedia – Social and Behavioral Sciences, 144*, 378–391. https://doi.org/10.1016/j.sbspro.2014.07.307
22. Bourdieu, P. (1986). The forms of capital. In J. Richardson (Ed.), *Handbook of theory and research for the sociology of education* (pp. 241–258). Greenwood.
23. Palhares, J. C. P., & Pezzopane, J. R. M. (2015). Water footprint accounting and scarcity indicators of conventional and organic dairy production systems. *Journal of Cleaner Production, 93*, 299–307. https://doi.org/10.1016/j.jclepro.2015.01.035
24. Parsa, H. G., & Narapareddy, V. (2015). *Sustainability, social responsibility, and innovations in the hospitality industry* (p. 370). Apple Academic Press. ISBN 9781926895673.
25. Borrini-Feyerabend, G., Kothari, A., & Oviedo, G. (2004). *Indigenous and local communities and protected areas: Towards equity and enhanced conservation* (p. xviii+111). IUCN.

26. Kelly, J., Haider, W., Williams, P. W., & Englund, K. (2007). Stated preferences of tourists for eco-efficient destination planning options. *Tourism Management, 28*(2), 377–390. https://doi.org/10.1016/j.tourman.2006.04.015
27. Han, H., et al. (2011). Are lodging customers ready to go green? An examination of attitudes, demographics, and eco-friendly intentions. *International Journal of Hospitality Management, 30*(2), 345–355. https://doi.org/10.1016/j.ijhm.2010.07.008
28. Han, H., Hsu, L., & Sheu, C. (2010). Application of the theory of planned behavior to green hotel choice: Testing the effect of environmental friendly activities. *Tourism Management, 31*(3), 325–334. https://doi.org/10.1016/j.tourman.2009.03.013
29. Johnson, D. (2002). Environmentally sustainable cruise tourism: A reality check. *Marine Policy, 26*(4), 261–270. https://doi.org/10.1016/S0308-597X(02)00008-8
30. Jørgensen, M. T., Hansen, A. V., Sørensen, F., Fuglsang, L., Sundbo, J., & Jensen, J. F. (2021). Collective tourism social entrepreneurship: A means for community mobilization and social transformation. *Annals of Tourism Research, 88*, 103171. https://doi.org/10.1016/j.annals.2021.103171
31. Okazaki, E. (2008). A community-based tourism model: Its conception and use. *Journal of Sustainable Tourism, 16*(5), 511–529. https://doi.org/10.1080/09669580802159594
32. Higham, J., Cohen, S. A., Cavaliere, C. T., Reis, A., & Finkler, W. (2016). Climate change, tourist air travel and radical emissions reduction. *Journal of Cleaner Production, 111*(Part B), 336–347. https://doi.org/10.1016/j.jclepro.2014.10.100
33. Chen, M.-F., & Tung, P.-J. (2014). Developing an extended theory of planned behavior model to predict consumers' intention to visit green hotels. *International Journal of Hospitality Management, 36*, 221–230. https://doi.org/10.1016/j.ijhm.2013.09.006
34. Huang, H.-C., Lin, T.-H., Lai, M.-C., & Lin, T.-L. (2014). Environmental consciousness and green customer behavior: An examination of motivation crowding effect. *International Journal of Hospitality Management, 40*, 139–149. https://doi.org/10.1016/j.ijhm.2014.04.006
35. Chen, R. J. C. (2015). From sustainability to customer loyalty: A case of full service hotels' guests. *Journal of Retailing and Consumer Services, 22*, 261–265. https://doi.org/10.1016/j.jretconser.2014.08.007
36. Law, A., De Lacy, T., Lipman, G., & Jiang, M. (2016). Transitioning to a green economy: The case of tourism in Bali, Indonesia. *Journal of Cleaner Production, 111*(Part B), 295–305. https://doi.org/10.1016/j.jclepro.2014.12.070
37. Han, H., & Yoon, H. J. (2015). Hotel customers' environmentally responsible behavioral intention: Impact of key constructs on decision in green consumerism. *International Journal of Hospitality Management, 45*, 22–33. https://doi.org/10.1016/j.ijhm.2014.11.004
38. Han, H. (2015). Travelers' pro-environmental behavior in a green lodging context: Converging value-belief-norm theory and the theory of planned behavior. *Tourism Management, 47*, 164–177. https://doi.org/10.1016/j.tourman.2014.09.014
39. Klein, R. A. (2011). Responsible cruise tourism: Issues of cruise tourism and sustainability. *Journal of Hospitality and Tourism Management, 18*(1), 107–116. https://doi.org/10.1375/jhtm.18.1.107
40. Niehm, L. S., Swinney, J., & Miller, N. J. (2008). Community social responsibility and its consequences for family business performance. *Journal of Small Business Management, 46*(3), 331–350. https://doi.org/10.1111/j.1540-627X.2008.00247.x
41. Palmer, T., & Riera, A. (2003). Tourism and environmental taxes. With special reference to the "Balearic ecotax". *Tourism Management, 24*(6), 665–674. https://doi.org/10.1016/S0261-5177(03)00046-3
42. Rodrigue, J. P., & Notteboom, T. (2013). The geography of cruises: Itineraries, not destinations. *Applied Geography, 38*, 31–42. https://doi.org/10.1016/j.apgeog.2012.11.011
43. Pásková, M. (2014a). Geoturismus v geoparcích v ČR a ve světě [Geotourism in the Czech and world geoparks]. In D. Milošová et al. (Eds.), *Modul průvodce ekoturismu a geoturismu [Module for ecotourism and geotourism guides]*. Gaudeamus Hradec Králové. ISBN 978-80-7435-330-7, 158 s.

44. Cohen, E. (1988b). Authenticity and commoditization in tourism. *Annals of Tourism Research, 15*(3), 371–386. https://doi.org/10.1016/0160-7383(88)90028-X
45. Pásková, M. (2014). *Udržitelnost cestovního ruchu [Tourism sustainability]* (p. 335). Gaudeamus Hradec Králové. ISBN 978-80-7435-329-1.
46. Vlachos, P. A., Tsamakos, A., Vrechopoulos, A. P., & Avramidis, P. K. (2009). Corporate social responsibility: Attributions, loyalty, and the mediating role of trust. *Journal of the Academy of Marketing Science, 37*(2), 170–180. https://doi.org/10.1007/s11747-008-0117-x
47. Young, M., Higham, J. E. S., & Reis, A. C. (2014). 'Up in the air': A conceptual critique of flying addiction. *Annals of Tourism Research, 49*, 51–64. https://doi.org/10.1016/j.annals.2014.08.003
48. Zahra, A., & McGehee, N. G. (2013). Volunteer tourism: A host community capital perspective. *Annals of Tourism Research, 42*, 22–45. https://doi.org/10.1016/j.annals.2013.01.008
49. Zelenka, J. (2001). Integrated management. *E+M Ekonomie a Management, 4*(1), 11–14.
50. Butler, R. W. (1980). The concept of a tourism area cycle of evolution: Implication for management of resources. *Canadian Geographer, 24*(1), 5–12. https://doi.org/10.1111/j.1541-0064.1980.tb00970.x
51. Munday, M., Turner, K., & Jones, C. (2013). Accounting for the carbon associated with regional tourism consumption. *Tourism Management, 36*, 35–44. https://doi.org/10.1016/j.tourman.2012.11.005
52. González-Rodríguez, M. R., Díaz-Fernández, M. C., & Simonetti, B. (2015). The social, economic and environmental dimensions of corporate social responsibility: The role played by consumers and potential entrepreneurs. *International Business Review, 24*(5), 836–848. https://doi.org/10.1016/j.ibusrev.2015.03.002
53. Lin, C.-H., Yang, H.-L., & Liou, D.-Y. (2009). The impact of corporate social responsibility on financial performance: Evidence from business in Taiwan. *Technology in Society, 31*(1), 56–63. https://doi.org/10.1016/j.techsoc.2008.10.004
54. López-Mosquera, N., & Sánchez, M. (2012). Theory of planned behavior and the value-belief-norm theory explaining willingness to pay for a suburban park. *Journal of Environmental Management, 113*, 251–262. https://doi.org/10.1016/j.jenvman.2012.08.029
55. Grayson, D., & Hodges, A. (2004). *Corporate social opportunity: 7 steps to make CSR work for your business* (p. 390). Greenleaf.
56. Liu, Z. (2003). Sustainable tourism development: A critique. *Journal of Sustainable Tourism, 11*(6), 459–475. https://doi.org/10.1080/09669580308667216
57. Asatryan, R. (2012). Modelling the cost-benefits of Corporate Social Responsibility (CSR) for decision making in the Aviation Industry. In *Advances in economics, risk management, political and law science, business and economics series* (pp. 108–113). WSEAS. [online]. Retrieved January 14, 2022, from http://www.wseas.us/e-library/conferences/2012/Zlin/EPRI/EPRI-16.pdf
58. UNWTO. (2022b). Global code of ethics for tourism, [online]. Retrieved February 04, 2022, from https://www.unwto.org/global-code-of-ethics-for-tourism
59. Young, A., & Kielkiewicz-Young, A. (2001). Sustainable supply network management. *Corporate Environmental Strategy, 8*(3), 260–268. https://doi.org/10.1016/S1066-7938(01)00122-1
60. UNWTO. (2013). Sustainable tourism for development guidebook (2013). Enhancing capacities for Sustainable Tourism for development in developing countries. UNWTO, EU Grant contract DCI-MULTI-2011/280-663, p. 228, [online], Retrieved August 14, 2014, from http://www.ilo.org/wcmsp5/groups/public/%2D%2D-ed_dialogue/%2D%2D-sector/documents/publication/wcms_216669.pdf
61. Becker-Olsen, K. L., Cudmore, B. A., & Hill, R. P. (2006). The impact of perceived corporate social responsibility on consumer behavior. *Journal of Business Research, 59*(1), 46–53. https://doi.org/10.1016/j.jbusres.2005.01.001
62. Argandoña, A. (2010). Corporate social responsibility in the tourism industry: Some lessons learned from the Spanish experience. IESE Business School Working Paper No. 844, [online]. Retrieved January 14, 2016, from http://core.ac.uk/download/files/153/6320426.pdf

63. Sheppard, D. (2006). The new paradigm for protected areas: Implications for managing visitors in protected areas. In D. Siegrist et al. (eds.) *Exploring the nature of management*. Proceedings of the Third International Conference on Monitoring and Management of Visitor Flows in Recreational and Protected Areas. Rapperswil, Switzerland, 13–17 September, pp. 33–45, ISBN 3-033-00935-2, [online]. Retrieved January 25, 2024, from https://mmv.boku.ac.at/refbase/files/sheppard_david-2006-the_new_paradigm_for.pdf
64. Eccles, R. G., & Stroehle, J. (2018). Exploring social origins in the construction of ESG Measures. https://ssrn.com/abstract=3212685 or https://doi.org/10.2139/ssrn.3212685
65. Lee, K. H., Packer, J., & Scott, N. (2015). Travel lifestyle preferences and destination activity choices of slow food members and non-members. *Tourism Management, 46*, 1–10. https://doi.org/10.1016/j.tourman.2014.05.008
66. Ionescu, G. H., Firoiu, D., Pirvu, R., & Vilag, R. D. (2019). The impact of ESG factors on market value of companies from travel and tourism industry. *Technological and Economic Development of Economy, 25*(5), 820–849. https://doi.org/10.3846/tede.2019.10294
67. Bonilla-Priego, J., Font, X., & Pacheco-Olivares, R. (2014). Corporate sustainability reporting index and baseline data for the cruise industry. *Tourism Management, 44*, 149–160. https://doi.org/10.1016/j.tourman.2014.03.004
68. Mancha, R. M., & Yoder, C. Y. (2015). Cultural antecedents of green behavioral intent: An environmental theory of planned behavior. *Journal of Environmental Psychology, 43*, 145–154. https://doi.org/10.1016/j.jenvp.2015.06.005
69. Marimon, F., Llach, J., & Bernardo, M. (2011). Comparative analysis of diffusion of the ISO 14001 standard by sector of activity. *Journal of Cleaner Production, 19*(15), 1734–1744. https://doi.org/10.1016/j.jclepro.2011.06.003
70. de Grosbois, D. (2012). Corporate social responsibility reporting by the global hotel industry: Commitment, initiatives and performance. *International Journal of Hospitality Management, 31*(3), 896–905. https://doi.org/10.1016/j.ijhm.2011.10.008
71. Gao, Y. L., & Mattila, A. S. (2014). Improving consumer satisfaction in green hotels: The roles of perceived warmth, perceived competence, and CSR motive. *International Journal of Hospitality Management, 42*, 20–31. https://doi.org/10.1016/j.ijhm.2014.06.003
72. Medina, L. K. (2005). Ecotourism and certification: Confronting the principles and pragmatics of socially responsible tourism. *Journal of Sustainable Tourism, 13*(3), 281–295. https://doi.org/10.1080/01434630508668557
73. Dawkins, J. (2004). Corporate responsibility: The communication challenge. *Journal of Communication Management, 9*(2), 108–119. https://doi.org/10.1108/13632540510621362
74. Font, X., et al. (2012). Corporate social responsibility: The disclosure–performance gap. *Tourism Management, 33*(6), 1544–1553. https://doi.org/10.1016/j.tourman.2012.02.012
75. Frey, N., & George, R. (2010). Responsible tourism management: The missing link between business owners' attitudes and behaviour in the Cape Town tourism industry. *Tourism Management, 31*(5), 621–628. https://doi.org/10.1016/j.tourman.2009.06.017
76. Janusz, G. K., & Bajdor, P. (2013). Towards to sustainable tourism – Framework, activities and dimensions. *Procedia Economics and Finance, 6*, 523–529. https://doi.org/10.1016/S2212-5671(13)00170-6
77. Holcomb, J. L., Upchurch, R. S., & Okumus, F. (2007). Corporate social responsibility: What are top hotel companies reporting? *International Journal of Contemporary Hospitality Management, 19*(6), 461–475. https://doi.org/10.1108/09596110710775129
78. WWF. (2022). Code of conduct for Arctic Tourists, WWF International Arctic Programme, [online]. Retrieved August 10, 2022, from https://wwfeu.awsassets.panda.org/downloads/codeofconductforarctictourists(eng).pdf
79. Sigala, M. (2008). A supply chain management approach for investigating the role of tour operators on sustainable tourism: The case of TUI. *Journal of Cleaner Production, 16*(15), 1589–1599. https://doi.org/10.1016/j.jclepro.2008.04.021
80. Jones, S. (2005). Community-based ecotourism: The significance of social capital. *Annals of Tourism Research, 32*(2), 303–324. https://doi.org/10.1016/j.annals.2004.06.007

81. Shaw, G., & Williams, A. M. (2002). *Critical issues in tourism: A geographical perspective* (p. 371). Blackwell Publishers.
82. Pearce, D. G. (1987). *Tourism today. A geographical analysis* (p. 229). University of Canterbury.
83. Chen, F.-Y., Chang, Y. H., & Lin, Y. H. (2012). Customer perceptions of airline social responsibility and its effect on loyalty. *Journal of Air Transport Management, 20*, 49–51. https://doi.org/10.1016/j.jairtraman.2011.11.007
84. Ciliberti, F., Pontrandolfo, P., & Scozzi, B. (2008). Investigating corporate social responsibility in supply chains: A SME perspective. *Journal of Cleaner Production, 16*(15), 1579–1588. https://doi.org/10.1016/j.jclepro.2008.04.016
85. Cochran, P. L. (2007). The evolution of corporate social responsibility. *Business Horizons, 50*(6), 449–454. https://doi.org/10.1016/j.bushor.2007.06.004
86. Falck, O., & Heblich, S. (2007). Corporate social responsibility: Doing well by doing good. *Business Horizons, 50*(3), 247–254. https://doi.org/10.1016/j.bushor.2006.12.002
87. Fu, H., Ye, B. H., & Law, R. (2014). You do well and I do well? The behavioral consequences of corporate social responsibility. *International Journal of Hospitality Management, 40*, 62–70. https://doi.org/10.1016/j.ijhm.2014.03.004
88. Garay, L., & Font, X. (2012). Doing good to do well? Corporate social responsibility reasons, practices and impacts in small and medium accommodation enterprises. *International Journal of Hospitality Management, 31*(2), 329–337. https://doi.org/10.1016/j.ijhm.2011.04.013
89. Inoue, Y., & Lee, S. (2011). Effects of different dimensions of corporate social responsibility on corporate financial performance in tourism-related industries. *Tourism Management, 32*(4), 790–804. https://doi.org/10.1016/j.tourman.2010.06.019
90. Joppe, M. (1996). Sustainable community tourism development revisited. *Tourism Management, 17*(7), 475–479. https://doi.org/10.1016/S0261-5177(96)00065-9
91. Mostafanezhad, M. (2014). Volunteer tourism and the popular humanitarian gaze. *Geoforum, 54*, 111–118. https://doi.org/10.1016/j.geoforum.2014.04.004
92. Weaver, D. B. (2012b). Organic, incremental and induced paths to sustainable mass tourism convergence. *Tourism Management, 33*(5), 1030–1037. https://doi.org/10.1016/j.tourman.2011.08.011
93. Weaver, A. (2005a). The Mcdonaldization thesis and cruise tourism. *Annals of Tourism Research, 32*(2), 346–366. https://doi.org/10.1016/j.annals.2004.07.005
94. Bauman, C. W., & Skitka, L. J. (2012). Corporate social responsibility as a source of employee satisfaction. *Research in Organizational Behaviour, 32*, 63–86. https://doi.org/10.1016/j.riob.2012.11.002
95. Farrell, B. H., & Twining-Ward, L. (2004). Reconceptualizing tourism. *Annals of Tourism Research, 31*(2), 274–295. https://doi.org/10.1016/j.annals.2003.12.002
96. Chan, E. S. W., & Hawkins, R. (2010). Attitude towards EMSs in an international hotel: An exploratory case study. *International Journal of Hospitality Management, 29*(4), 641–651. https://doi.org/10.1016/j.ijhm.2009.12.002
97. Chan, E. S. W., et al. (2014). What drives employees' intentions to implement green practices in hotels? The role of knowledge, awareness, concern and ecological behaviour. *International Journal of Hospitality Management, 40*, 20–28. https://doi.org/10.1016/j.ijhm.2014.03.001
98. Farooq, M., Farooq, O., & Jasimuddin, S. M. (2014). Employees response to corporate social responsibility: Exploring the role of employees' collectivist orientation. *European Management Journal, 32*(6), 916–927. https://doi.org/10.1016/j.emj.2014.03.002
99. Sin, H. L. (2010). Who are we responsible to? Locals' tales of volunteer tourism. *Geoforum, 41*(6), 983–992. https://doi.org/10.1016/j.geoforum.2010.08.007
100. WTA. (2007). Code of conduct – Wildlife Tourism Australia, [online]. Retrieved June 22, 2017, from http://www.wildlifetourism.org.au/about/guidelines/code-of-conduct/
101. Arıkan, E., & Güner, S. (2013). The impact of corporate social responsibility, service quality and customer-company identification on customers. *Procedia – Social and Behavioral Sciences, 99*, 304–313. https://doi.org/10.1016/j.sbspro.2013.10.498

102. LastFrontiers. (2016). Code of ethics for travellers, [online]. Retrieved April 10, 2016, from http://www.lastfrontiers.com/about-us/responsible-tourism/last-frontiers-code-of-ethics-for-travellers
103. Morrison, A. P. (1995). *Hospitality and travel marketing* (p. 567). Cengage Learning, ISBN 0-8273-6620-5.
104. Fisher, B., Turner, R. K., & Morling, P. (2009). Defining and classifying ecosystem services for decision making. *Ecological Economics, 68*(3), 643–653. https://doi.org/10.1016/j.ecolecon.2008.09.014
105. European Commission. (2005). A manual for evaluating the quality performance of tourist destinations and services, Directorate-General for Enterprise and Industry, Publications Office, p. 60, [online]. Retrieved August 19, 2022, from https://op.europa.eu/en/publication-detail/-/publication/d6a1c97f-3080-4eed-8e22-d30766542ab6
106. Fair Trade. (2022). Fairtrade international, [online]. Retrieved August 14, 2022, from https://www.fairtrade.net/
107. European Parliament. (2013b). REPORT on Corporate Social Responsibility: Promoting society's interests and a route to sustainable and inclusive recovery (2012/2097(INI)), [online]. Retrieved November 14, 2016, from https://www.europarl.europa.eu/doceo/document/A-7-2013-0023_EN.html
108. European Parliament. (2014). Directive 2014/95/EU of the European Parliament and of the Council of 22 October 2014 amending Directive 2013/34/EU as regards disclosure of non-financial and diversity information by certain large undertakings and groups Text with EEA relevance, [online]. Retrieved January 14, 2016, from https://eur-lex.europa.eu/legal-content/EN/TXT/?uri=CELEX:32014L0095
109. Basterrechea, C. (2015). Environmental commitment. A responsible company in the tourism industry, [online]. Retrieved February 14, 2016, from http://www.fondazioneuniverde.it/wp-content/uploads/2015/02/Presentazione-NH-Hotel-Group.compressed.pdf
110. Chiu, Y.-T. H., Lee, W.-I., & Chen, T.-H. (2014). Environmentally responsible behavior in ecotourism: Antecedents and implications. *Tourism Management, 40*, 321–329. https://doi.org/10.1016/j.tourman.2013.06.013
111. Dwyer, L., et al. (2009). Destination and enterprise management for a tourism future. *Tourism Management, 30*(1), 63–74. https://doi.org/10.1016/j.tourman.2008.04.002
112. Choi, H. C., & Sirakaya, E. (2006). Sustainability indicators for managing community tourism. *Tourism Management, 27*(6), 1274–1289. https://doi.org/10.1016/j.tourman.2005.05.018
113. Tsaur, S. H., Lin, Y.-C., & Lin, J.-H. (2006). Evaluating ecotourism sustainability from the integrated perspective of resource, community and tourism. *Tourism Management, 27*(4), 640–653. https://doi.org/10.1016/j.tourman.2005.02.006
114. Cucculelli, M., & Goffi, G. (2016). Does sustainability enhance tourism destination competitiveness? Evidence from Italian Destinations of Excellence. *Journal of Cleaner Production, 111*(Part B), 370–382. https://doi.org/10.1016/j.jclepro.2014.12.069
115. Jha, A., & Cox, J. (2015). Corporate social responsibility and social capital. *Journal of Banking & Finance, 60*, 252–270. https://doi.org/10.1016/j.jbankfin.2015.08.003
116. Thompson, M. H., Dumont, C. P., & Gaymer, C. F. (2008). ISO 14001: Towards international quality environmental management standards for marine protected areas. *Ocean & Coastal Management, 51*(11), 727–739. https://doi.org/10.1016/j.ocecoaman.2008.07.004
117. Franzoni, S. (2015). Measuring the sustainability performance of the tourism sector. *Tourism Management Perspectives, 16*, 22–27. https://doi.org/10.1016/j.tmp.2015.05.007
118. Hsiao, T.-Y., et al. (2014). Establishing attributes of an environmental management system for green hotel evaluation. *International Journal of Hospitality Management, 36*, 197–208. https://doi.org/10.1016/j.ijhm.2013.09.005
119. Hunter, C., & Shaw, J. (2007). The ecological footprint as a key indicator of sustainable tourism. *Tourism Management, 28*(1), 46–57. https://doi.org/10.1016/j.tourman.2005.07.016
120. Gössling, S., et al. (2011). Food management in tourism: Reducing tourism's carbon 'foodprint'. *Tourism Management, 32*(3), 534–543. https://doi.org/10.1016/j.tourman.2010.04.006

121. Gössling, S., & Buckley, R. (2016). Carbon labels in tourism: Persuasive communication? *Journal of Cleaner Production, 111*(Part B), 358–369. https://doi.org/10.1016/j.jclepro.2014.08.067
122. Miossec, J. M. (1976). *Elements pour une Theorie de l'Espace Touristique. Les Cahiers du Tourisme* (p. 63). Centre des Hautes Études Touristiques, Aix-en-Provence.
123. Waligo, V. M., Clarke, J., & Hawkins, R. (2013). Implementing sustainable tourism: A multi-stakeholder involvement management framework. *Tourism Management, 36*, 342–353. https://doi.org/10.1016/j.tourman.2012.10.008
124. Gallardo-Vázquez, D., & Sanchez-Hernandez, M. I. (2014). Measuring corporate social responsibility for competitive success at a regional level. *Journal of Cleaner Production, 72*, 14–22. https://doi.org/10.1016/j.jclepro.2014.02.051
125. Hein, L., van Koppen, C. S. A., van Ierland, E. C., & Leidekker, J. (2016). Temporal scales, ecosystem dynamics, stakeholders and the valuation of ecosystems services. *Ecosystem Services, 21*(Part A), 109–119. https://doi.org/10.1016/j.ecoser.2016.07.008
126. Kerala. (2008). The Kerala declaration on responsible tourism, second international conference on responsible tourism in destinations, [online]. Retrieved February 09, 2016, from http://www.responsibletourism2008.org/keraladeclaration.php
127. Campbell, E. T., & Brown, M. T. (2012). Environmental accounting of natural capital and ecosystem services for the US National Forest System. *Environment, Development and Sustainability, 14*(5), 691–724. https://doi.org/10.1007/s10668-012-9348-6
128. Garrod, B., & Fyall, A. (1998). Beyond the rhetoric of sustainable tourism? *Tourism Management, 19*(3), 199–212. https://doi.org/10.1016/S0261-5177(98)00013-2
129. Rahman, I., Reynolds, D., & Svaren, S. (2012). How "green" are North American hotels? An exploration of low-cost adoption practices. *International Journal of Hospitality Management, 31*(3), 720–727. https://doi.org/10.1016/j.ijhm.2011.09.008
130. Gössling, S., et al. (2005). The eco-efficiency of tourism. *Ecological Economics, 54*(4), 417–434. https://doi.org/10.1016/j.ecolecon.2004.10.006
131. Koubská, K., & Hralová, E. (2006). Společensky odpovědné podnikání jako trend a příležitost [Socially responsible business as a trend and opportunity], Centre for Innovation and Development, p. 44, [online]. Retrieved May 21, 2015, from www.cir.cz/prirucka-csr/482649/1985622
132. Benavides-Velasco, C. A., Quintana-García, C., & Marchante-Lara, M. (2014). Total quality management, corporate social responsibility and performance in the hotel industry. *International Journal of Hospitality Management, 41*, 77–87. https://doi.org/10.1016/j.ijhm.2014.05.003
133. UNWTO. (1992). Guidelines: Development of national parks and protected areas for tourism, p. 53, ISBN 92-844-0026-0.
134. Lindberg, K., Enriquez, J., & Sproule, K. (1996). Ecotourism questioned: Case studies from Belize. *Annals of Tourism Research, 23*(3), 543–562. https://doi.org/10.1016/0160-7383(95)00074-7
135. British Airways. (2022). British Airways – Responsible business, [online]. Retrieved July 22, 2022, from https://www.britishairways.com/en-gb/information/about-ba/ba-better-world/responsible-business
136. Campbell, C. L., Heinrich, D., & Schoenmüller, V. (2015). Consumers' reaction to fair trade motivated price increases. *Journal of Retailing and Consumer Services, 24*, 79–84. https://doi.org/10.1016/j.jretconser.2015.02.005
137. Kettl, K.-H. (2011). Ecological footprint comparison of different means of transportation based on Sustainable Process Index (Spi) methodology. In K.-H. Kettl (Ed.), *Proceedings of 1st international conference on sustainable intelligent manufacturing*. June 28–July 1.
138. Kolk, A. (2016). The social responsibility of international business: From ethics and the environment to CSR and sustainable development. *Journal of World Business, 51*(1), 23–34. https://doi.org/10.1016/j.jwb.2015.08.010
139. Windsor, D. (2013). Corporate social responsibility and irresponsibility: A positive theory approach. *Journal of Business Research, 66*(10), 1937–1944. https://doi.org/10.1016/j.jbusres.2013.02.016

140. Pásková, M. (2012). Environmentalistika cestovního ruchu [Environmental tourism]. *Czech Journal of Tourism, 1*(2), 77–113.
141. Pásková, M. (2017). Local and indigenous knowledge regarding the land use and use of other natural resources in the Aspiring Rio Coco Geopark. In *World Multidisciplinary Earth Sciences Symposium (WMESS 2017) IOP Publishing IOP Conf. Series: Earth and Environmental Science* (Vol. 95, p. 052018). IOP Publishing. https://doi.org/10.1088/1755-1315/95/5/052018
142. easyJet. (2015). Presentation of Corporate Social Responsibility on the easyJet, plc website in 2015, [online]. Retrieved August 21, 2015, from http://corporate.easyjet.com/corporate-responsibility.aspx?sc_lang=en
143. Yunis. (2006). Making tourism more sustainable. Helpful instruments and practice examples in Europa. Prezentace 3. 2. 2006, Hannover, 19. s., [online]. Retrieved August 14, 2014, from http://www.ecotrans.org/docs/1_hamele_Yunis_intro_aims.pdf
144. Tejedo, P., Benayas, J., Cajiao, D., Leung, Y.-F., De Filippo, D., & Liggett, D. (2022). What are the real environmental impacts of Antarctic tourism? Unveiling their importance through a comprehensive meta-analysis. *Journal of Environmental Management, 308*, 14634. https://doi.org/10.1016/j.jenvman.2022.114634
145. Buckley, R. (2002). Tourism ecolabels. *Annals of Tourism Research, 29*(1), 183–208. https://doi.org/10.1016/S0160-7383(01)00035-4
146. Commission of the European Communities. (2003). Basic orientations for the sustainability of European tourism, COM(2003) 716 final. https://eur-lex.europa.eu/LexUriServ/LexUriServ.do?uri=COM:2003:0716:FIN:en:PDF
147. Erkuş-Öztürk, H., & ERAYDIN, A. (2010). Environmental governance for sustainable tourism development: Collaborative networks and organisation building in the Antalya tourism region. *Tourism Management, 31*(1), 113–124. https://doi.org/10.1016/j.tourman.2009.01.002
148. Kumar, D. (2022). Economic and political uncertainties and sustainability disclosures in the tourism sector firms. *Tourism Economics, 29*(6), 1694–1699. https://doi.org/10.1177/13548166221113434
149. Singal, M. (2014). Corporate social responsibility in the hospitality and tourism industry: Do family control and financial condition matter? *International Journal of Hospitality Management, 36*, 81–89. https://doi.org/10.1016/j.ijhm.2013.08.002
150. Lei, T., Wei, Y., & Otilia, M. (2022). Artificial Intelligence influences intelligent automation in tourism: A mediating role of internet of things and environmental, social, and governance investment. *Frontiers in Environmental Science, 10*, 853302. https://doi.org/10.3389/fenvs.2022.853302
151. Fletcher, R., Mas, I. M., Blanco-Romero, A., & Blázquez-Salom, M. (2019). Tourism and degrowth: An emerging agenda for research and praxis. *Journal of Sustainable Tourism, 27*(12), 1745–1763. https://doi.org/10.1080/09669582.2019.1679822
152. Henderson, J. C. (2007a). Corporate social responsibility and tourism: Hotel companies in Phuket, Thailand, after the Indian Ocean tsunami. *International Journal of Hospitality Management, 26*(1), 228–239. https://doi.org/10.1016/j.ijhm.2006.02.001
153. Budeanu, A. (2005). Impacts and responsibilities for sustainable tourism: A tour operator's perspective. *Journal of Cleaner Production, 13*(2), 89–97. https://doi.org/10.1016/j.jclepro.2003.12.024
154. TIAC. (2022). Code of ethics and guidelines for sustainable tourism, tourism ndustry Association of Canada and National Round Table on the Environment and the Economy, [online]. Retrieved August 14, 2022, from http://nrt-trn.ca/wp-content/uploads/2011/08/sustainable-tourism.pdf
155. TIA. (2016). Code of ethics – The Tourism Industry Association New Zealand (TIA), [online]. Retrieved April 10, 2016, from http://www.tianz.org.nz/main/code-of-ethics/
156. Wokoun, R., & Vystoupil, J. (1987). *Geografie cestovního ruchu a rekreace I [Geography of tourism and recreation I]* (p. 250). Státní pedagogické nakladatelství.
157. Font, X. (2002). Environmental certification in tourism and hospitality: Progress, process and prospects. *Tourism Management, 23*(3), 197–205. https://doi.org/10.1016/S0261-5177(01)00084-X

158. Sharma, A., Moon, J., & Strohbehn, C. (2014). Restaurant's decision to purchase local foods: Influence of value chain activities. *International Journal of Hospitality Management, 39*, 130–143. https://doi.org/10.1016/j.ijhm.2014.01.009
159. Lin, L.-Y., Tsai, C.-C., & Lee, J.-Y. (2022). A study on the trends of the global cruise tourism industry, sustainable development, and the impacts of the COVID-19 pandemic. *Sustainability, 14*(11), 6890. https://doi.org/10.3390/su14116890
160. Henderson, J. C. (2007b). *Tourism crises: Causes, consequences and management* (p. 186). Routledge.

Chapter 4
Social Responsibility in Tourism and Its Specific Features

Tourism is a sector that requires a high degree of shared responsibility and integration in its implementation. The distinctive nature of social responsibility in tourism is mainly due to the large number of actors (Fig. 4.1) that should be engaged in its definition, declaration and implementation. Social responsibility is very closely related to sustainable tourism and its potential implementation is intricately linked to the past development of tourism. If social responsibility in tourism is not applied effectively, many undesirable effects happen to tourism destinations:

- The conditions for the existence and further development of tourism in the destination are deteriorating, which manifests itself in a reduction in its originality and uniqueness, deterioration in the quality of the environment, a change in the attitude of the local community towards tourism and other changes in the destination that lead to a reduction in its attractiveness, quality of tourist experience and demand.
- The undesirable impact of tourism on other sectors in the destination multiplies (negative multiplier effect), which is particularly significant in the consumption of destination resources (e.g. undeveloped space, water and agricultural land) and environmental pollution.

Implementing social responsibility in tourism requires extensive interdisciplinary collaboration and careful consideration of the needs of both local individuals and the community as a whole. Should the long-term sustainability of tourism be ensured, both individual and collective tourism actors must embrace social responsibility, as shown schematically in Fig. 4.1. It is evident that the effectiveness of social responsibility in tourism is also conditioned by the fact that a large number of tourism actors are involved. The effect of social responsibility in a destination can be significantly degraded just by a few tourism actors with very low levels of social responsibility. For example, visitors deliberately damaging natural and cultural heritage, or engaging in high-impact outdoor activities such as off-road motorcycling and in other irresponsible outdoor activities. These visitors are often unaware of

M. Pásková, J. Zelenka, *Social Responsibility in Tourism*, Tourism on the Verge,
https://doi.org/10.1007/978-3-031-61609-9_4

Fig. 4.1 Collective and individual actors of social responsibility in tourism

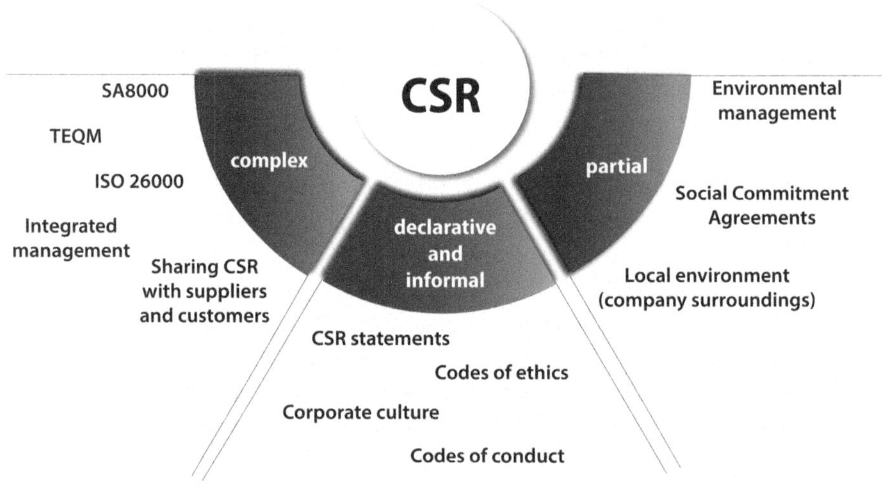

Fig. 4.2 Ways of implementing corporate social responsibility

their irresponsibility because they do not consider the consequences of their behaviour.

Figure 4.2 shows how the implementation of corporate social responsibility takes various forms, not only in the tourism sector. If the inclusion of different dimensions of social responsibility is sufficiently holistic, this implementation can be **complex**.

Such a holistic approach is supported by formal and standardised system solutions of company management (ISO 26000, SA 8000), gradually standardised system solutions of company management in the form of integrated management, social responsibility shared with suppliers (supply-customer chains), customers/participants of tourism, and informal system solutions (e.g. TQM and TQEM). According to ISO, integrated management is implemented as a combination of ISO 9001+, ISO 14001+ and ISO 45001:2018.

Partial approaches focus only on the environmental dimension, such as environmental management in the organisation—ISO 14000+, or commitments to care for the local environment, or on socio-cultural and economic responsibility like agreements on social commitments to the local community, such as preferential employment of local residents.

Declarative and informal approaches are, alone or in combination with other approaches, a way of implementing social responsibility. They are expressed in codes of ethics, codes of conduct, social responsibility statements and the corporate culture of the organisation.

4.1 Theoretical Foundations of Responsible Tourism in Relation to Social Responsibility

The foundations of the theoretical concept of responsible tourism in relation to social responsibility are:

- Social exchange theory.
- The concept of sustainable development is based on planning, management and balancing of three or more aspects of development, most often environmental, economic and socio-cultural.
- Approaches based on the participatory management and social capital development.
- Knowledge management with systematic use of current knowledge of tourism processes, including the possibility of their sufficiently accurate conceptual and factual prediction[1]—examples include modelling of tourism processes in [1, 2, 3] and the use of the destination life cycle concept in Pásková [4, 5].

Social exchange theory (see [4, 5]) describes and explains a situation where, in an attempt to achieve balance in the guest–host relationship, values damaged by certain social groups such as tourism operators and tourism participants are compensated to other groups like local communities in the destination, source and transit areas or, in the best case, no harm is done due to prevention.

The theoretical basis for increasing social responsibility of tourism actors is relevant interdisciplinary studies, such as social ecology (understanding the social causes of environmental problems, including understanding the reasons for the

[1] Modelling, neural networks, extrapolation etc.

emergence of negative externalities in tourism and the reasons for the unsustainability of tourism development), environmental philosophy (application to mountain resorts, e.g. [6]) and environmental economics.[2]

Other concepts relevant to analysing various aspects of social responsibility in relation to responsible tourism include:

- **Life cycle analysis** applied to:
  - Tourism products in relation to the degree of consideration of social responsibility in all phases of product development.
  - Tourism destinations in relation to tourism sustainability and the application of the type and degree of social responsibility in all stages of destination development ([4, 5, 7]).
  - Sustainability of destination development with the possibility to change the existing tourism in the destination to a more responsible/sustainable tourism based on a higher level of social responsibility and the resulting paradigm shift ([8, 9, 10, 11]).
- **Causal analysis**, conducted as an analysis of interactions between society and the environment, referred to as DPSIR (Driving forces, Pressures, States, Impacts, Responses; [12, 13, 14]). As shown in Fig. 4.3, a general conceptual analysis of social responsibility in sustainable tourism using the DPSIR method can causally verify the strengths and weaknesses, and possibly opportunities and threats, of the application of social responsibility in relation to tourism trends, motivations and behaviours of tourism actors and phenomena in a particular destination. It can thus be used to develop a SWOT analysis or other methods of analysing the state of a particular entity or destination.
- **Indicative analysis of the impacts of tourism and the resulting state of the environment** in relation to the declared and achieved level of social responsibility of organisations (e.g. airlines and hotel chains) in their behaviour and implementation of processes with environmental impact. This type of analysis uses in particular:
  - Calculation of the aggregate **ecological footprint**, representing the total area required to cover the consumption of resources (energy, water etc.) and deposition of waste over a specific period of time for an individual, certain types of activities or local communities. Modified/simplified sub-footprints focusing on a specific phenomenon or consequence can also be used to indicate impact. This is, for example the carbon footprint, which expresses the amount of CO_2 generated directly and indirectly, as shown by Cadarso et al. [15] by calculating the total carbon footprint, including, in addition to direct emissions, emissions associated with tourism infrastructure investments.

[2] Economisation of aspects not yet or insufficiently considered by the market—externalities, ecosystem services etc.

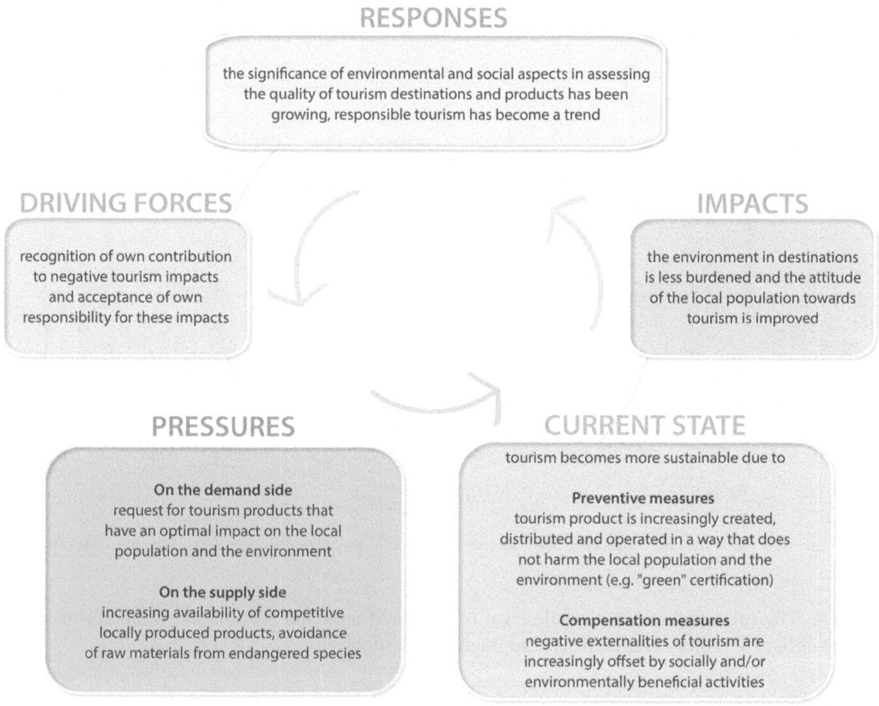

Fig. 4.3 DPSIR method applied for depiction of the causal process of tourism responsibility

- **Tourism sustainability indicators** ([16, 17]), namely general indicators applicable to all types of destinations and destination-specific indicators.
- **LAC method** (Limits of Acceptable Change); [18, 19, 20].
- **Carrying capacity** with its multiple dimensions such as psychological, socio-cultural or ecological carrying capacity and finding the relationship between the tourism pressure on the area and the subsequent impacts of tourism in the area [1, 4, 5, 7, 21, 22, 23, 24, 25, 26].

• **An analysis of negative and positive externalities** of tourism, see, for example Pásková and Zelenka [27].
• **The assessment of ecosystem services** ([28, 29, 30]), which is the basis for the economic valuation of benefits given by nature to humans, including those for tourism.
• **Respecting and managing the limits of destination development** as expressed by:

- **Strategic planning and strategic management.**
- **Carrying capacity.**
- **The Limits of Acceptable Change** (LAC) **method.**
- **Tourism sustainability indicators** ([16, 17, 31]).

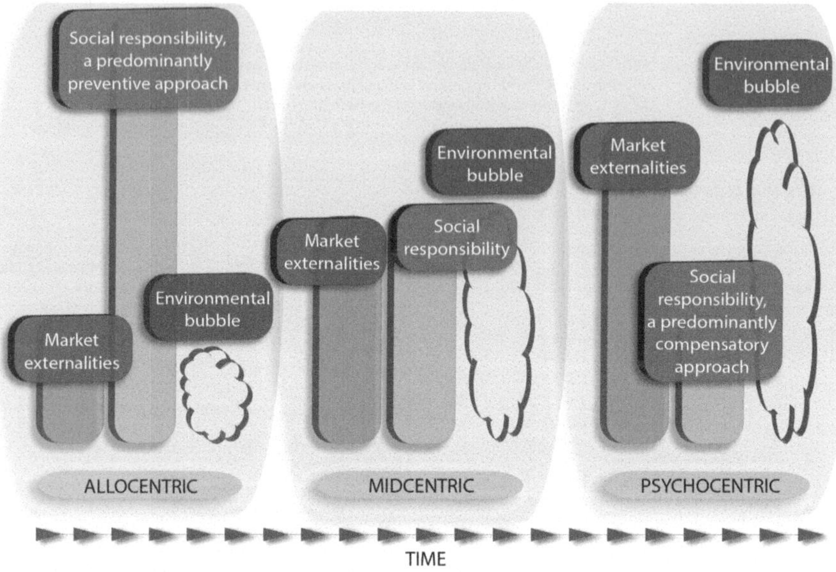

Fig. 4.4 The relationship between the size of the environmental bubble and the level of social responsibility of visitors and their psychographic classification

- **Visitor management models** ([20, 32]) are typically based on the preservation of the visitor experience or the LAC method.

- **Psychographics** (typically applied in marketing) helps distinguish visitor segments according to their underlying motivations, behavioural patterns and degree of social responsibility (Fig. 4.4; [33]). Visitors' behavioural patterns can be observed and recorded using spatio-temporal visitor detection.
- **The ethics of tourism**, which is the basis for both formal and informal social responsibilities of various tourism actors (e.g. [34, 35]). As Caton [36] argues, so far little attention has been paid to the moral aspects of tourism in relation to their importance in tourism studies (e.g. [37]).

4.2 Socio-cultural, Environmental and Economic Dimensions of Responsibility in Tourism

In Chap. 2, especially in Fig. 2.2, the key aspects that determine sustainable tourism under the assumption of mutually balanced development have been presented, emphasising and classifying the three basic dimensions of sustainability—socio-cultural, environmental and economic. Comprehensive use of ICT is an important factor that can significantly support socially responsible development in tourism, as illustrated by the reasoning and practical examples of Jucan and Baier [38], Nicolae and Sabina [39], Zelenka et al. [2] and Zelenka, Pásková and Husáková [3]. According to them, ICT as an important component of knowledge management can significantly increase the access to information and knowledge necessary for responsible decision-making for all tourism stakeholders. ICT supports the environmental and economic efficiency of processes and systems[3] and enables better competitiveness of organisations that adopt social responsibility by means of marketing communication with tourism stakeholders.

In this chapter, the socio-cultural, environmental and economic dimensions of tourism responsibility are considered in the context of civil society, society's attitude to nature and cost-effectiveness.

4.2.1 Civil Society as a Basis for Social Responsibility

In order for society as a whole to maximise benefits and minimise undesirable environmental and community impacts of its consumption behaviour on the environment and the global community, it is essential that its individual members are aware of these impacts and understand the effects of their feedback. In this respect, the quality of the education system, education and training is crucial. Another prerequisite for responsible social behaviour is to consider society's value system, in particular, the positioning of the quality of the environment and the relationship with nature, with one's own and other communities, in the collective value system of the society. In terms of the acceptance and application of the principles of social responsibility by individual citizens, the setting of the institutional environment—the existence of relevant rules, such as norms, standards, certifications, ecolabelling, unwritten rules, public institutions and incentive mechanisms, including the recognition and rewarding of examples of good practice—plays a crucial role. The features and dynamics of the institutional environment are largely influenced by the importance of the community and its collective memory for individual members of society—in the case of tourism, for the residents of the destination. An engaged, responsible and accountable civil society creates social and economic demand for an environmentally and socially responsible approach and output from companies and other business and public entities.

Tourism participants differ from each other in terms of their degree of social responsibility depending on their psychographic classification ([33]) and the corresponding size of the environmental bubble (Fig. 4.4). Using the necessary generalisation, it is a continuum of visitors:

- A psychocentric with a typically large environmental bubble showing little or zero social responsibility.

[3] For example, the impact of satellite navigation in aviation and higher capacity utilisation due to reservation systems.

- A midcentric with an average environmental bubble and a limited degree of social responsibility.
- An allocentric with a small environmental bubble and a relatively high level of social responsibility.[4]

The degree of social responsibility of visitors changes during the life cycle of a destination in relation to changes in the quality and quantity of services provided and the degree of its touristification (for details, see Plog 1977; Pásková 2003, 2012, 2014). During certain stages of the destination's development, when tourists who prefer familiar and comfortable environments become the dominant group, it becomes crucial for the long-term viability of tourism to enhance the efforts of destination management and individual tourism businesses in practicing social responsibility.

Similarly to the environmental dimension, the approach to addressing the socio-cultural dimension of responsibility can be divided into three main directions according to the attitude towards responsibility:

a) **Socio-cultural compensation measures** may include, for example financial support for carrying out local traditions and activities of associations, or privileged access to visitor infrastructure for local residents. Ideally, these measures compensate for specific consequences of negative externalities caused to the destination communities by the activities of tourism operators and by the presence and activities of visitors.
b) **Avoidance of negative externalities** is based, for example on the application of the concept of fair trade, see Sect. 4.7, or on the preferential employment of local people and systematic support for the improvement of their skills.
c) **Creation of positive externalities** can promote the socially desirable development of the region. According to Poria et al. [40], examples include the responsibility of tourism operators in Israel towards the local community to stabilise the social situation, and the introduction of environmentally friendly transportation and other environmentally friendly technologies to the destination. In this way, both socio-cultural and environmental dimensions of social responsibility are addressed.

4.2.2 Society's Attitude Towards Nature as a Basis for Social Responsibility

An essential prerequisite for a more significant application of social responsibility within its environmental dimension is positioning of nature higher on the individual and collective value ladder. This is linked to an awareness of the functioning and value of ecosystems and a strong territorial identity.

[4] i.e. high respect for local communities and their authentic lifestyle and environment.

The dimension of environmental responsibility can be carried out in three basic ways:

a) **Environmental compensation measures** include afforestation or river revitalisation. Ideally, these measures remedy specific consequences of negative externalities that affect the environment of the destination as a result of the activities of a tourism operator. In tourism, a typical example is airline offset programmes (Sect. 3.5).
b) **Avoidance of negative externalities** includes regulation of the intensity and spatio-temporal distribution of tourism, especially in naturally and socio-culturally vulnerable locations, and application of green management or relevant ISO standards in a hotel or hotel chain. A typical precautionary approach to regulating tourism activities may be, for example to close a tourist trail where a protected species is being reintroduced (e.g. nesting of the peregrine falcon in Broumovsko Protected Landscape Area; [41]). In such cases, however, it is necessary to consider and possibly model in advance ([2]) the so-called opportunity costs of such a measure. These may take the form, for example of shifting visitor flows from a closed trail to another trail, which may lead to increased environmental impacts in the vicinity of the latter.
c) **Creation of positive externalities** may include, for example the introduction of environmentally friendly transport to places frequented by local residents, or the cleaning of water bodies used by local residents.

4.2.3 Cost-Effectiveness as a Basis for Social Responsibility

Awareness of market failures and the competitive environment, such as market externalities, unequal access to information and resources, along with a long-term approach that includes planning and evaluating environmental investments and innovations, are prerequisites for the practical application of the economic dimension of social responsibility.

Again, there are three main ways of implementing the economic dimension of responsibility:

a) **Economic compensation measures** ideally remedy specific consequences of negative economic externalities caused by the tourism operator. These measures include the revitalisation of local brownfields, financial support for teaching or marketing local crafts, traditional agriculture etc.
b) **Prevention of negative externalities** includes, but is not limited to, consistent application of participatory management, preference for local suppliers of services and goods,[5] and preferential employment of local residents in tourism services. A precautionary approach also includes regulation of tourism activities in

[5] For example, restaurants preferring to buy local food [42].

cultural and historical destinations or tourist attractions with reduced costs for maintenance, repairs and tourism infrastructure.

c) **Creation of positive externalities**, multiplication of income, employment and investments made through tourism in other sectors of the local economy including agriculture, local transport, small crafts and services such as hairdressing and repair of sports equipment.

4.2.4 Environmental Behaviour of Socially Responsible Tourism Providers

So far, the most important part of the precautionary approaches to social responsibility is the reduction of its negative impact on the environment. The following chapters analyse the state of the art and provide typical approaches and examples of reducing these impacts in air travel, cruise shipping, hotels and catering.

Air Travel

Tourism-related air transport has seen continuous growth in recent years and its impact on the environment has also been increasing. This is linked to CO_2 and NO_x emissions in the high atmosphere, and thus to global warming and the Earth's ozone layer, as well as noise pollution and, to a lesser extent, light pollution in the vicinity of airports. The sector tends to be associated with high fuel consumption per unit of transport capacity (passenger-kilometre) and high overall emissions. It is important to compare its relative consumption figures with those of other passenger transport modes. Detailed analyses using the Sustainable Process Index (SPI), which comprehensively assesses emissions using the ecological footprint concept ([43]), show much higher impacts of passenger car transport (with the exception of biogas cars) compared to air transport, and conversely significantly lower emissions from trains and buses in this comparison.

In recent years, new technologies have been developed and introduced in air transport quite successfully. Their typical feature is the balance between economic interests and the reduction of negative impacts on the environment. In particular, cost savings due to reduced fuel consumption per passenger resulting from higher aircraft capacity, more efficient power units and better aerodynamics. Technically, this has been achieved in recent years by modifying wingtips with winglets, aerodynamically shaping engines and reducing aircraft dead weight (see, e.g. [44]), as well as higher flight occupancy due to competition, high-quality central and global reservation systems and the use of Yield Management approaches.

An interesting counter-intuitive finding is provided by Arjomandi and Seufert [45], who analysed how environmental attitudes and performance differ between traditional full-service and low-cost airlines. Surprisingly, the results of their comparative study show that low-cost airlines are more environmentally oriented by the nature of their low-cost strategy. Factors include newer and more homogeneous aircraft fleets, higher aircraft occupancy, less waste due to non-catering etc. The environmental orientation of traditional airlines has also been growing gradually and is also used in their marketing (e.g. [46]).

Traditional airlines, in particular, have been gradually adopting system solutions, especially the introduction of Environmental Management Systems (EMS) according to ISO 14001+ and EMAS, or simpler methods of cleaner production. An example is Scandinavian Airlines' application of EMS according to ISO 14001+, which Lynes and Andrachuk [47] analysed including the motivation for social responsibility. According to Quazi et al. [48], the motivations for EMS adoption are cost savings, top management interest, employee satisfaction, compliance with environmental regulations, meeting customer expectations, concerns about business barriers, respect for the environmental practices of corporate management and gaining a competitive advantage. The International Air Transport Association (IATA) is active in promoting EMS and other approaches to reducing or offsetting negative environmental impacts. This organisation offers environmental audits to airlines ([49]) and promotes the marketing-friendly implementation of greenhouse gas (GHG) offset programmes—the "Carbon Offset Program" ([50, 51, 52, 53]). An essential prospective sub-measure is the promotion of the use of alternative fuels (currently mainly biofuels), which IATA also supports ([49, 54]).

All these approaches are presented on the airlines' websites, social media accounts, airport terminals and on-board infotainment systems in line with the creation of an attractive airline image. Indeed, according to studies by Hagmann, Semeijn and Vellenga [55] and Mayer, Ryley and Gillingwater [56], the image of airlines, which includes the perception of a favourable relationship between the airline and the environment, positively influences the purchasing behaviour of potential customers, who are willing to pay extra for an 'environmentally friendly' alternative, even if less than for extra services. However, the results of studies related to willingness to pay (e.g. [57]) need to be taken with caution. Similar to the difference between the stated content and scope of social responsibility and the actual practice of many tourism operators, the reality of customers paying extra for an "environmentally friendly" choice[6] is different from their declared behaviour.

ESG effectiveness in air travel is supported by favourable external conditions, where airlines based in countries with comprehensive support for business and ICT use demonstrate superior environmental, social and governance performance compared to airlines based in countries with high government influence on the aviation sector ([58]).

[6] For example, a premium to offset their carbon footprint on a flight.

Collective Tourist Accommodation

Hotels and hotel chains are showing several different trends, sometimes even contradictory ones. A significant trend is the increasing complexity and quality of services in response to the growing real or perceived environmental bubble (see Fig. 4.4, a discussion of the issue in the context of tourism sustainability in [5]) of hotel guests and the associated higher demands for quality and complexity of services in hotels and hotel resorts. Changes in the quality and scope of services are also based on competitive practices to achieve a certain local or even global uniqueness ("to differentiate, to be the biggest, the highest, the most luxurious, the most spectacular, the most modern…"). This is confirmed by Whitla, Walters and Davies [59] who consider the emphasis on marketing and competitiveness as the basis of global strategy in the hotel industry. These trends contribute to negative socio-economic phenomena such as the emergence of dual worlds or the reduction of diversity and endogeneity of the local economy. At the same time, however, there are also trends that have a positive impact on the environment, in particular, the reduction of energy consumption and the need for accommodation facilities to exploit other natural resources.

When accommodation facilities use local resources, including the employment of local people, there is a positive impact on local communities and their economy, for example by supporting local farmers, artisans and local food manufacturers, in the construction and supply of accommodation facilities. Accommodation establishments can also provide targeted support to local suppliers, such as organic farmers and local employment by providing education and training for them. This brings educational and social benefits to the local population. These positive trends are based on changes in visitor preferences when searching for and choosing accommodation. Hotel management also takes into account the declared societal demands as stated in strategic and/or conceptual materials created at different geographical levels (including, for example the EU, see Antošová, Csikósová 2015). Whitla, Walters and Davies (2007) demonstrate that customers are prepared to share social responsibility with accommodation providers and expect this responsibility in their services.

Similar to airlines, the increase in social responsibility of hotels manifests itself gradually through the introduction of technical and technological measures that help find a balance between cost savings, environmental impacts and comfort. An example is the replacement of part or all capacity of conventional air conditioning with high-quality building insulation and the heat recovery system that provides the basis for the concept of passive houses or low-energy buildings (e.g. [60]). Benchmarking is used to demonstrate the increased efficiency of these measures. A general review of the use of benchmarking in services improving resource efficiency was presented by Grove [61] and a comparison of water and energy consumptions using benchmarking for the Hilton Hotels and Resorts hotel chain was presented by Bohdanowicz and Martinac [62]. In addition to green management (see Sect. 4.8.2), which is implemented internationally mainly according to the rules of Green Key certification, hotels and hotel chains are also introducing Environmental Management

Systems (EMS) according to international standards ISO 14001+ (e.g. [63, 64]). In hotels, green management is certified under the Green Key label [65]. According to Černec [66], the hotel chains Best Western, IHG, NH Hotels and Carlson Residor have implemented green management. Criteria and methodologies for implementing green management are also issued for campsites, small accommodation facilities, restaurants and tourist attractions.

According to Quazi et al. [48] and Chan and Wong [67], the main motivating factors for EMS implementation in the form of ISO 14001+ in the hotel industry are mainly internal factors—corporate governance, ISO benefits and the certainty of compliance with environmental legislation, whilst emphasising the important role of top management decisions. They also list external motivating factors—market trends, trade barriers and customers. On the other hand, according to Chan [63], the main barriers to EMS are the lack of knowledge and skills, lack of expert advice, uncertainty in achieving the desired outcome, concerns about the requirements for certification, lack of resources and the cost of EMS implementation and maintenance. Chan [64, 68] analyses the nature of changes achieved by the implementation of ISO 14001+ in large hotels and states that half of the measures were related to energy savings. He also describes ways of evaluating the energy consumption of hotels[7] and a comparative method of energy regimes for comparing hotels.

Crucial to the prosperity of hotels that have implemented green management is their perception by guests (e.g. [69]), the subsequent behavioural responses of guests, the immediate and long-term popularity of such hotels and the willingness of guests to pay the same or higher price compared to other hotels (e.g. [70]). Tilikidou, Delistavrou and Sapountzis [71] observe that guests are willing to stay longer in hotels with green management but do not actively seek them out, and these hotels had higher support amongst guests with higher education.

Chan and Hawkins [72] analysed the EMS implementation in the hotel industry and concluded that the most important factors are public support in the early stage, implementation of pilot validation, preliminary problem analysis and collaboration with an external consulting firm. In the subsequent phases, in accordance with the ISO 14001, it is important to establish an environmental policy,[8] planning, internal and external communication of the results achieved (i.e. progress in fulfilling social responsibility), control of processes, training and increasing the competence of employees and continuous improvement, which also means increasing social responsibility.

Ayala [73] summarises the environmental approaches used in the hotel industry: eco-techniques, environmental sponsorship and the creation of environmental service packages. He also discusses a new trend, namely the creation of "eco-resorts" in relation to the principles of ecotourism.

In many cases, eco-resorts can be more of a marketing-created "pseudo-attraction" for visitors, as their consumption of energy and other resources

[7] In rooms, cooking, heating bathing water etc.

[8] That is, to declare the content of social responsibility in relation to the environment.

(typically water) is often similar to that of traditional hotel and other accommodation facilities, as Warnken, Bradley and Guilding [74] demonstrate for Australia.

Cruise Ships on Seas and Oceans

The offer of cruises originated in the 1960s as a response to the insurmountable competition of air transport in transoceanic transport. For that reason, special vessels were created for pleasure trips. According to Rodrigue and Notteboom (2013), these cruises have become a very clear symbol of the globalisation of the tourism industry in the way they reach and cover the market, the massive scale and focus of their services, and the mobility of their assets. Geographically, they report year-round averaged cruise performance, particularly in the Caribbean (43% of passengers) and the Mediterranean (23% of passengers).

Carić [75] observed several trends in the cruise ship segment, leading to an increase in air emissions, and associated in recent years with an increase in the size of newly built ships, a reduction in port calls, and a significant increase in fleet size and passenger numbers. Specifically, as Rodrigue and Notteboom [76] commented on these trends, the average stopover time in the Caribbean in 2013 was 4.3 h, and according to CLIA [77], as cited in Klein [78], the cruise segment has grown 7.2% annually since 1990, making it the fastest growing segment of the tourism industry. This growth has sparked discussions and studies (e.g. [79] focusing on the Arctic) and sometimes real activities (Antarctica) towards regional regulation of cruise ships. At the same time, Carić [75] points out that the negative consequences are particularly pronounced in developing countries and transitional economies, which relates to the geographical distribution of ship destinations (see above) as well as their underdeveloped waste management. In another study, Carić and Mackelworth [80] describe the potential consequences of cruises in the Adriatic Sea and approaches to regulate them.

Cruise ships threaten the environment mainly by emitting CO_2, various types of polluted water and solid waste ([78]). As reported by Rodrigue and Notteboom [76], CO_2 emissions per passenger per day can vary widely by ship type, size, age and amenities, and are significantly lower for large mass tourism ships (up to 1200x) compared to small luxury ships (Fig. 4.5). Klein [78] notes that these effects are very significant and, for example are several orders of magnitude higher for cruise ships using cheap fuel due to the large addition of sulphur in this fuel compared to automotive fuel per litre.

Butt [81] analyses the negative implications of waste disposal generated on cruise ships for home and frequently visited ports and emphasises the need to implement quality waste management in ports, including recycling, reuse and prevention of waste generation on ships. The desired reduction of these impacts creates a large potential for the introduction of various social responsibility approaches. Today, the growth of cruises can be perceived as a typical example of globalisation ([82]), McDonaldization ([83]) and inherently unsustainable tourism with adverse consequences in terms of increased resource consumption, environmental pollution and significant socio-cultural impacts in the visited destinations. Johnson [84] therefore stressed that adequate social responsibility and close cooperation between the management of cruise companies and the management of the destinations visited by

Fig. 4.5 A small luxury boat on the Paraguay River in the Pantanal, the world's largest tropical wetland area, is an example of enormous resource consumption and a lack of social responsibility. Photo: J. Zelenka, 2008

cruise passengers is necessary in order to increase the sustainability of cruises. He recommends focusing on waste management and developing intergenerational and internal social capital rather than short-term economic gain.

The results of the study of the preferences of cruise participants as presented by Adams [85] at the Innovators Symposium in Grenada are relevant to the marketing of cruise businesses in the context of social responsibility:

1. Cruise participants prefer companies that implement social responsibility.
2. The most important factors in their decision-making are the quality and price provided, followed by the cruise itinerary with a description of the destinations and the length of the cruise. If these parameters are comparable, these passengers also consider social responsibility.
3. Upper-class passengers are more socially responsible and are willing to pay more for socially responsible products.
4. Frequent cruise ship clients prefer companies that implement social responsibility. They are therefore able to influence the implementation of social responsibility more than other passengers.
5. Considering the increasing size of ships and the focus of marketing communication on younger passengers, it can be expected that social responsibility will remain of marginal importance for cruise ships.

Items 3 and 4 of the findings presented in Adams [85] confirm the general observation implied by other studies (e.g. [86]) that higher-income travellers can more easily afford to pay for their own moral comfort and social responsibility. This applies, however, when comparing people whose relationship to the environment is approximately of the same position in their value ranking (Ceteris paribus).

The SARS-COV-2 pandemic had a very significant impact on the ocean cruise sector ([87]). The first studies analysing this impact (e.g. [87, 88]), suggest not only the adoption of more vigorous prevention of the spread of contagious diseases ([87]) but also a greater emphasis on lower environmental impacts, especially by using different propulsion units, as well as a discontinuation of the trend to build ever larger ships.

Catering Facilities

The effects of activities in restaurants on perceptions of restaurant quality and customer loyalty in three types of restaurants (high-end, mid-range and fast-food restaurants) were examined by Namkung and Jang [89]. The first type of activity focuses on food origin (organic and locally sourced) and the second one is dedicated to other environmental management of operations, especially on energy efficiency and recycling. Their results show the differences in customer perceptions for different types of restaurants. For higher-end restaurants, perceptions of the "green brand" were associated with the environmental aspects of the food, especially the origin of the food, rather than other environmental aspects of the operation. For mid-range restaurants, the opposite was true, and for fast-food restaurants there was no demonstrable influence of either approach. In all types of restaurants, a growing segment of diners showed the expected positive correlation between the degree to which diners prefer environmentally friendly approaches and the degree to which they perceive higher quality in restaurants that apply these approaches. Campbell and DiPietro [90] supplemented their findings on guest consumption behaviour with the assertion that guest interest in local products and produce increases when they can identify with the region through photographs and other representations of local produce, products and farms. Guest preferences for organic products were investigated by Dowd and Burke [91] who found that health considerations and ethical values were the most important decision factors. They also confirmed the importance of ethical aspects, especially the promotion of regional production, quality of animal life and good prices for farmers. Their research was based on the application of the theory of planned holiday.

4.3 Motivating Tourism Actors to Act in a Socially Responsible Way

Research on the motivations for embracing socially responsible behaviour is very important for understanding the behaviour of tourism actors and their social responsibility performance. As many studies have shown (e.g. [92]), besides external legislative pressures and perceived client preferences, the economic benefit of introducing social responsibility in firms is also an important motivator for the implementation of social responsibility. The expectation of positive economic returns resulting from the adoption of CSR is supported by many studies. For example, Rodríguez and Cruz [93] and Saeidi et al. [94] document a significant positive correlation between CSR adoption and firm productivity. According to Saeidi et al. [94], corporate social responsibility has a positive impact on an organisation's competitiveness and reputation and on customer satisfaction.

Even the same types of tourism actors have different motivations running their businesses, by which these actors can be further segmented. This is illustrated, for example by Font, Garay and Jones [95] research on small tourism businesses in European protected areas, which identified three groups of companies:

- Business- and profit-oriented companies that engage mainly in environmental efficiency.
- Companies motivated by legislative and social conformity, including in relation to tourism actors, engaged in a wide range of social responsibility activities.
- Companies based on lifestyle and moral values, engaged in a large number of environmental, social and economic activities related to social responsibility.

These results show the importance of appropriate motivation for the efficiency and success of applying CSR approaches and for its declared and actual contents.

As Jha and Cox [96] point out, the motivation for embracing social responsibility in a destination can also be the high value of social capital in the destination, which is a very important finding for tourism, its sustainability and the development of socially responsible activities and forms of tourism. Boosting social responsibility in a destination also increases its social capital. Indeed, social responsibility motivates tourism actors to identify common values for mutual cooperation and the generated positive externalities further motivate the growth of social capital. Font, Garay and Jones [95] confirmed this positive correlation between social responsibility and social capital by finding that companies performing social responsibility in a region with high social capital are also characterised by a higher degree of social responsibility.

4.4 EU Policy, Codes of Ethics, Charters, Codes of Conduct and International Standards

Due to the nature of tourism, which consists of the activities of many differently motivated tourism actors, the formal declaration of desirable ways of thinking and behaving in the form of transnational and national tourism policies, codes of ethics, charters and codes of conduct can be an important motivation and inspiration for a gradual change in their values' system and performance. This mental and behavioural status includes the relationship of tourism actors to nature and society and their cooperation with each other.

4.4.1 EU Policy

In recent years, the European Union has pursued a consistent policy of sustainable development. This is reflected in various declarations, directives, charters and thematic strategies (water, soil, landscape, cities etc.) in many areas of human activity, including tourism. The basic orientation of tourism development is expressed, inter alia, in the communications from the European Commission titled Basic Orientation for Sustainable European Tourism ([97]), Agenda for Sustainable and Competitive European Tourism ([98]), Europe, the World's No. 1 Tourist Destination—a new political framework for tourism in Europe ([99]) and Transition pathway for tourism ([100]). The last document declares ([100]:1): "*The aim of this report is to describe the measures and outputs needed to accelerate the green and digital transitions and improve the resilience of the tourism ecosystem*". In addition to the policy of sustainable and responsible development in tourism, detailed procedures for monitoring the sustainability of tourism are gradually being introduced within the EU. For destination management, ETIS (European Tourism Indicators System, European Commission 2016) was created. Various competitions (e.g. EDEN, see Sect. 5.3) and environmentally oriented certification schemes (e.g. Blue Flag, see Sect. 4.10; European Flower, ECEAT (European Centre for Ecology and Tourism) Quality Label and biohotels) are promoted to encourage desirable forms of tourism and destination management approaches. Environmental aspects and social responsibility are also briefly mentioned in the Hotelstars Union certification scheme for accommodation facilities, which is gradually being promoted in several European countries.

The attention paid to the sustainability of tourism is evidenced by the findings of the analyses of appropriate practices and approaches leading to its improvement, e.g. the final report of the Tourism Sustainability Group (TSG) of February 2007 ([101]). In the same way as geoparks cooperate within their national and continental networks and the global network (Sect. 6.1), the Network of European Regions for Competitive and Sustainable Tourism ([102]) is emerging. The aim is to create a future model of sustainable and competitive tourism, which is an important factor in the sustainable development of the region. It uses the concept of a combination of five key strategic approaches ([102]), known as the "*Five S: Smart destinations, Sociocultural balance, Skills and talent, Safety and resilience and Statistics and measurability*".

4.4.2 Charters

Charters are public proclamations of fundamental values, attitudes, concepts, rights or principles that can be voluntarily adhered to, even if they are not legally binding. In the last few decades, charters have often included an explicit commitment to responsible human attitudes towards the environment (both natural and anthropogenic), the biosphere and various parts of human society.

From a global as well as a European perspective, a number of charters are relevant to the sustainability of tourism and the responsibility of tourism actors—e.g. ICOMOS (International Council on Monuments and Sites) International Cultural Tourism Charter and Sites, the European Urban Charter and the Charter of European Planning. The World Conference on Sustainable Tourism's Charter for Sustainable Tourism ([64]), declared in 1995, is an important globally accepted charter. The basic premises, principles and objectives of the social responsibility in tourism contained in this Charter are shown in the mind map in Fig. 4.6.

Social responsibility in tourism is expressed not only in global and macro-regional charters (typically in Europe charters) but also in national, regional and even local tourism charters. Such charters are typically the expression of a long-term national or destination strategy focused on sustainable/responsible tourism. The following examples show the declaration of charters for two types of geographical levels (country, municipality), for protected areas and for geoparks.

Morocco is an example of a country that officially presents itself as a responsible tourism destination and this responsibility is expressed in the very comprehensive Moroccan Charter on Sustainable Tourism ([103]). Its content is in full compliance with the needs of responsible tourism and the World Charter for Sustainable

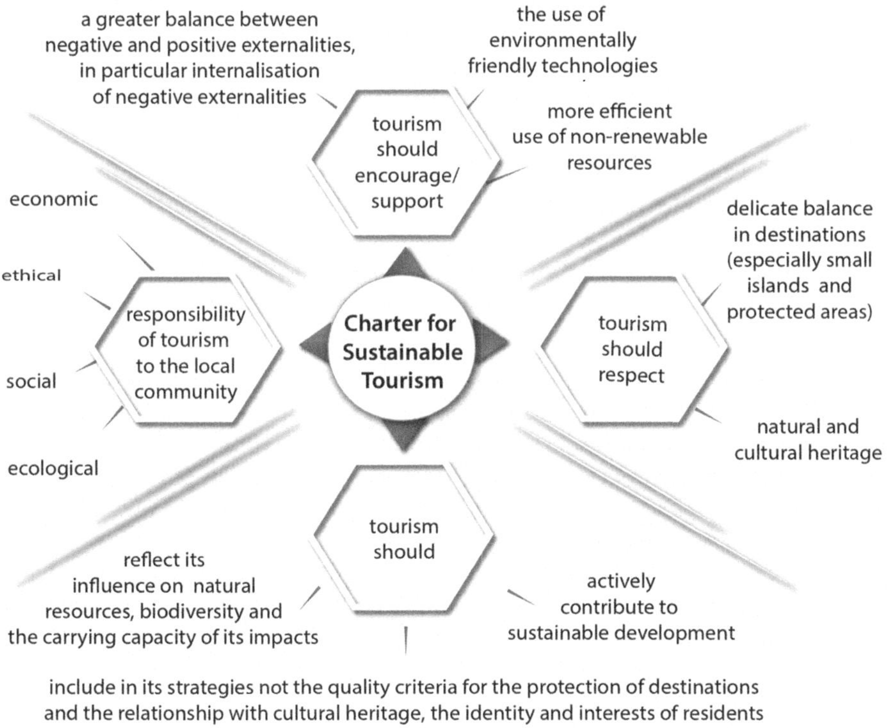

Fig. 4.6 A mind map of social responsibility, as expressed in the World Conference on Sustainable Tourism's Charter for Sustainable Tourism. Source: adapted from [64]

Tourism. It is, therefore, a voluntary collective commitment of local tourism operators and expresses, amongst other things, their respect for local culture, traditions, local people and, in particular, children and the environment.

An example of a municipal-level destination charter is the Cape Town Declaration on Responsible Tourism ([104]). This charter, despite its brevity, is very advanced in its vision and declares the participation of local stakeholders in tourism in a responsible manner, the adherence to important global documents[9] and the development of a policy and implementation plan for responsible tourism. It aims to optimise the impact of tourism, the quality of the visitor experience, the economic benefits for local people and their participation in tourism, and to stimulate subcontractors of tourism service providers to responsible business behaviour.

The European Charter for Sustainable Tourism in Protected Areas

The European Charter for Sustainable Tourism in Protected Areas is based on Agenda 21, adopted at the Earth Summit in Rio de Janeiro in 1992. The Charter is part of the Sixth Environment Action Programme and the EU Sustainable Development Strategy. The EUROPARC Federation, the umbrella organisation for protected areas in Europe, supervises the implementation of the mission of this document. An important tool for the implementation of the Charter can be the introduction of integrated tourism sustainability indicators, as described by Castellani and Sala [31]. According to the EUROPARC Federation [108]:4, the underlying aims of the charter are to:

- Increase awareness of, and support for, Europe's protected areas as a fundamental part of Earth heritage, that should be preserved for, and enjoyed by, current and future generations.
- Improve the sustainable development and management of tourism in protected areas, which takes account of the needs of the environment, local residents, local businesses and visitors.

The charter principles involve working in partnership, therefore it requires above all the application of cooperative management and participative management in the planning and management of tourism development ([108]:4): "*protected-area authorities should not work alone in the management of tourism, but rather that all those affected by tourism in and around a protected area should be involved in its development and management*".

According to the EUROPARC Federation [108]:4, the implementation of the Charter takes the form of transparent contracts and cooperation between protected area authorities, tourism service providers and local people, the charter "*encourages*

[9] Notably the Global Code of Ethics for Tourism ([105]), the Cape Town Declaration on Responsible Tourism in Destinations ([106]) and the Kerala Declaration on Responsible Tourism ([107]).

the sharing of responsibilities and outlines the individual and collective commitments of the protected area and its partners".

A fundamental commitment of the charter is the implementation of strategic tourism planning, which means "*making a diagnosis; consulting and involving partners; establishing strategic objectives; allocating the appropriate resources; following through an action plan, and evaluating the results*". ([108]:5).

The charter is divided into three sections ([108]):

I. **Sustainable tourism for the protected areas**

Protected area authorities are required to carry out a situational analysis of the protected area as a tourism destination in a participatory manner, i.e. with the involvement of all key tourism stakeholders, and to develop proposals for a vision and strategy for its further development.

II. **Sustainable tourism for local tourism businesses**

Tourism businesses must analyse the effects of their operations and the services they offer on customers, local people, the environment and cultural and natural heritage. Then, they have to propose ways to make tourism in the protected area more sustainable and how to cooperate with the protected area management as well as a monitoring system for control of its implementation.

III. **Sustainable tourism for companies organising tours**

Travel agencies, whose products are offered in the protected area with whose management and local residents they cooperate, identify the degree of compatibility of their objectives with the objectives of the protected area management and ensure that the increase in intensity and changes in the quality of tourism caused by their products do not result in damage to primary tourism resources, especially local natural and cultural heritage.

A particular protected area may accede to any of the three sections of the charter, depending on the interests of the entities concerned. The process of accession to the charter takes the form of a paid certification and the fulfilment of the declared parts of the charter is checked by EUROPARC inspectors.

European Charter of Geoparks

This charter ([109]), in various variations, is the fundamental ethical and philosophical document of geoparks, certified both globally ([110]) and nationally, such is for the Czech Republic Council of National Geoparks ([111]). It contains the basic principles and conditions for the establishment, development and functioning of geoparks ([112]). Its main principles include the active involvement of local residents in the geopark's activities and the promotion and education in geology and other Earth science disciplines. It also includes the issue of protection and maintenance of geopark sites, not in a legislative sense, but as an active engagement in education, economics, field engineering and other fields. It also expresses the collective commitment of geopark management and their strategic partners to zero tolerance for the destruction and damage of geological sites and, in the context of global geoparks, for the sale of precious stones and fossils. This is often due to the unknown or unproven origin of these natural resources and therefore uncertainty as to whether their extraction, processing and distribution has resulted in damage to nature and violations of human rights. The assumption of responsibility by the geopark's management takes place by the act of signing the charter, and a fundamental breach of its principles is grounds for withdrawal of the "UNESCO Global Geopark" or "National Geopark" certificate. These principles constitute a preventive form of social responsibility.

4.4.3 Codes of Ethics

Codes of ethics are a voluntary tool to increase the sustainability of tourism (Eurico, Oliveira and Hoefnagels [113]). Their extensive analysis was offered by Malloy and Fennell [114]. In relation to social responsibility, they documented that examined codes of ethics focused mainly on environmental and social aspects and only 9% of clauses in the codes were economically focused. This shows that codes of ethics are still insufficiently linked to the three-component concept of social responsibility and the concept of sustainable development. Another inspiration for the further development of codes of ethics brings Crang's [115] critical analysis of codes of ethics based on Western culture (Euro-American paradigm) in comparison with codes of ethics based on environmental ethics and the harmony of people with nature in Chinese culture. He thereby suggested that traditional Chinese culture can provide a much deeper basis for social responsibility. The deeper study of the Indigenous knowledge would probably bring the same finings on the unwritten codes of ethics of the Indigenous peoples regarding their both individual and collective perception of responsibility for the natural resources, perceived as Mother Earth ([116]).

In tourism, codes of ethics are adopted globally ([105, 117]), at the national level they are adopted by tourism business associations (e.g. [118, 119]), individual tourism businesses (e.g. LastFrontiers 2016), destination managements, state and local governments under the auspices of national tourism hubs, tourism ministries (e.g. Canada—[120]) as well as tourism destinations (e.g. North West Australia—[121]).

Global Code of Ethics for Tourism

The Global Code of Ethics for Tourism was approved by the UNWTO General Assembly in 1999 ([105]) and recognised by the UN General Assembly in 2001 ([113]). It is a code of ethics appealing to responsible behaviour and the corresponding mindset of tourism actors (especially visitors, local residents, tourism entrepreneurs and state and public administration) by defining the role of tourism in the world and an explicit list of appropriate practices for its sustainable development, for example promoting nature-based tourism and ecotourism and assessing the environmental impact of tourism investments. Within the scope of corporate social responsibility, the following are particularly relevant:

- **Principle of tolerance in attitudes**—respect for different religious, philosophical and moral backgrounds, opinions and beliefs and social and cultural traditions and practices.
- **Preventive dimension of responsibility**—preparing tourism actors to encounter different lifestyles, cultures and customs of both visitors and host communities.
- **Responsibility to respect human rights and the law**—basic knowledge of and respect for the relevant country's legislation, gender equality, the rights of vulnerable groups and minorities.
- **Responsibility for the protection of the environment and resources**, emphasising planning, protection of areas with particularly vulnerable ecosystems (e.g. promoting specific indicators for tourism sustainability; [122]).

The requirements of the Global Code of Ethics for Tourism are presented in more detail in the mind map (Fig. 4.7), developed on the basis of social responsibility in relation to tourism practice.

Codes of Ethics as Applied by Tourism Entities

Tourism organisations are increasingly introducing codes of ethics as a component of their corporate culture and social responsibility, which can be part of their communication with other tourism actors. In addition to declaring codes of ethics for their own employees and tourism participants, they also often communicate the codes of ethics valid in the destination (e.g. hotels in Guatemala communicate a code of ethics for visitors to Guatemala, see [123]). Eurico, Oliveira and Hoefnagels [113] identified three different forms of codes of ethics based on their research on hotel chains and family hotels in Europe:

- Codes of ethics similar in concept to codes of conduct, with a list of desirable behaviour and practices (e.g. [124]).
- Briefly stated basic ethical principles and values (e.g. [125]).
- Elaborate codes of ethics, covering also the relationship of hotel employees and hotel management to other tourism stakeholders (e.g. [126]).

The impact of codes of ethics on the internal environment of the company has been investigated by Lee et al. [127]. Their results show that a code of ethics has a direct effect on corporate philanthropy and social commitment of an organisation. According to this research, corporate philanthropy also positively influences the involvement of individual employees in corporate social responsibility initiatives.

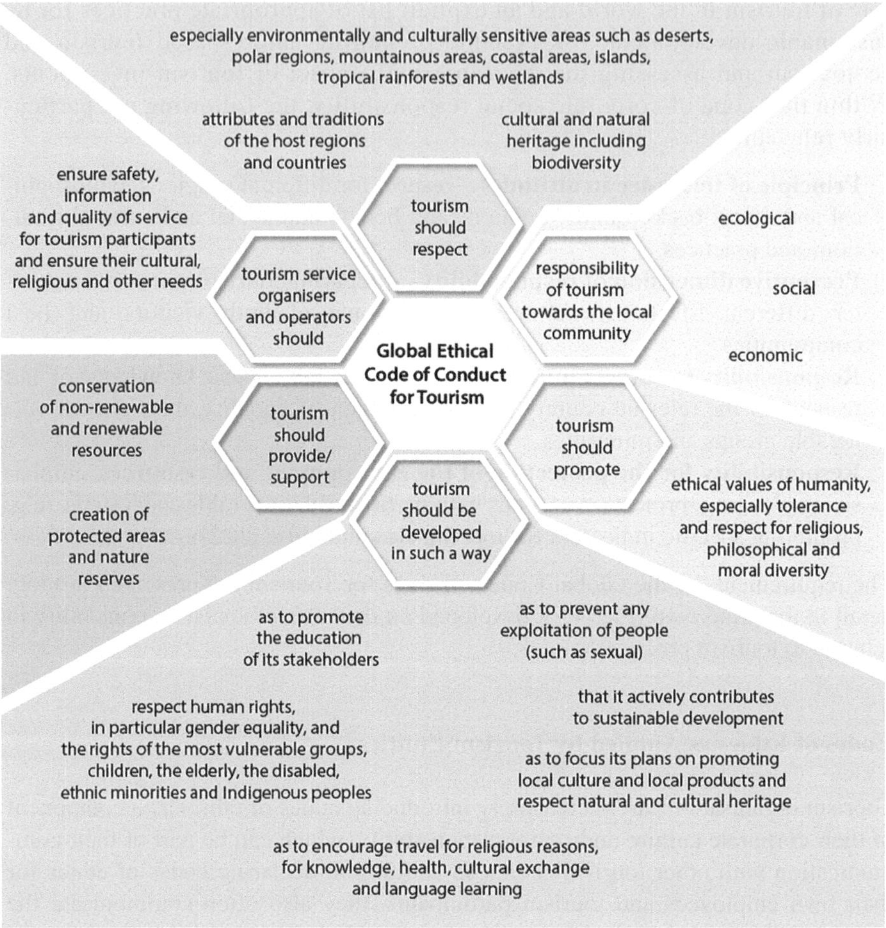

Fig. 4.7 Mind map of the content of social responsibility as it is contained in the Global Code of Ethics for Tourism. Source: Adapted from [105]

4.4.4 Codes of Conduct

Codes of conduct are more or less complex sets of recommended or even enforceable rules of conduct (e.g. the rules for visitors, transportation rules) for management and employees of tourism service providers and intermediaries, consumers of tourism services, including visitors to destinations. Some of these codes cover different aspects of social responsibility, as the case studies in the following subchapters show. At the same time, the content of codes of conduct may include, in addition to social responsibility, the desirable way of performing other activities, such as ensuring personal or information security. Some of these other activities may become a very important part of social responsibility in the near future, e.g.

ensuring the personal safety of visitors in relation to terrorism, war conflict and local crime. Similar to codes of ethics, codes of conduct are declared at different geographical levels. At the global level, for example the codes declared by ECPAT ([128]) for the protection of children against sexual exploitation in travel and tourism and ICNF ([129]) for nature tourism companies can be mentioned. At the macro-regional or continental level, there is, for example the Code of Conduct for Arctic tourists ([130]) or for tourists in Antarctica ([131]). At the national level, there are codes of conduct in use, e.g. Chile ([132]) or individual destinations such as Québec, Canada ([133]). Tour operators also draw up their own codes of conduct, for example the Code of Conduct for Yukon's Wilderness Tourism Operators ([134]) and similarly there is the Code of Conduct, Wildlife Tourism Australia Inc. ([135]).

Codes of Conduct in Accommodation Facilities

Codes of conduct and codes of ethics are mainly adopted by hotel chains (e.g. [136, 137]). Some hotel chains do not even formally distinguish between a code of ethics and a code of conduct ([137]). The scope of activities addressed by codes of conduct tends to be larger compared to activities currently included in corporate social responsibility (e.g. [136]). The code of conduct can apply not only to the employees of the accommodation facility itself but also to its guests and its suppliers (e.g. [138]).

Visitor Regulations

Many tourist attraction management bodies use visitor regulations for their own sustainability. In some tourist attraction areas, violating these regulations is sanctioned, e.g. some national parks or other types of protected areas. Elsewhere, visitors are guided on how to behave towards the environment, how to respect nature (Fig. 4.8), landscape, local people and cultural heritage. A visitor, tourism service provider, organiser or intermediary respecting the destination voluntarily assumes the ethical obligation to comply with the applicable rules and regulations. This contributes to improving the quality of visitor experience and supports tourism sustainability in the destination.

Unwritten Codes of Conduct

Visitors to some destinations, especially those with significant Indigenous populations, follow unwritten rules that are mediated by Indigenous peoples themselves, often acting as guides (see Sect. 5.5). Their perception of the Earth as sacred implies reverence, respect and a sense of belonging to nature. This attitude, a good knowledge of local ecosystems, local plants and animal species, is the basis for a responsible attitude towards natural resources. Ecosystems or populations are always used by them only as long as they can reproduce naturally. This ancient empirical

Fig. 4.8 Visitor regulations, both written and unwritten (and passed on by guides), are important for the conservation of endemic species such as lama vicuña (*Vicugna vicugna*), Colca y Volcanes de Andagua UNESCO Global Geopark, Peru. Photo: M. Pásková, 2015

recognition of carrying capacity allows Indigenous peoples to exercise responsibility in the use of life-sustaining natural resources without knowing theoretical concepts of "modern" science.

4.4.5 International Standards

Since the late 1980s, international standardisation has played an increasing role in the gradual promotion of appropriate sustainable development practices in different parts of society. In the following chapters, the use of international standards in relation to different environmental management practices (Sect. 4.8) and to the implementation of corporate social responsibility (Sects. 4.5 and 4.6) is presented.

The development of standards specific to tourism services and other tourism activities has also been very beneficial. These standards have gradually included both partial (e.g. requiring only an environmental policy) and comprehensive approaches to the sustainability of tourism. In addition to basic services, they also include very specific services, as can be seen from several examples of international ISO standards:

- In addition to the general standard for the provision of services in the tourism sector ([139]), the standard for a typical form of mass tourism—the provision of services in beach tourism ([140])—contains a comprehensive concept of how to carry out activities in line with sustainable tourism.
- ISO/TS 13811:2015 ([141]) responds to the significant environmental impact of accommodation facilities by seeking to optimise this impact.
- The requirements to develop an environmental policy are part of the standards for Thalassotherapy ([142]) and Spa Wellness ([143]).
- In natural protected areas, the ISO 18065:2015 ([144]) standard requires appropriate tourism infrastructure, waste management and environmental education.
- ISO for adventure tourism ([35]) and ISO for recreational diving ([96]) were also issued.

4.5 ISO 26000

ISO 26000 is a voluntary international standard that applies to all types of tourism organisations. It explains the complexity of CSR, supports the translation of CSR intentions into specific activities including best practice procedures and the appropriate sharing of results of its implementation with other tourism actors.

The adoption of the ISO 26000 standard in economic activities brings many different internal and external benefits to tourism organisations, both in the short and long term. This not only follows from the ISO standard itself, but also from the literature focused on its interpretation, e.g. in terms of its content and purpose (e.g. [145]), or from its links to other corporate responsibility approaches. These links between ISO 26000 and international integrated reporting are discussed, for example in [146]. According to ISO standards ([145, 146]), consistent application of CSR brings tourism organisations:

- Improving awareness of them, including improving their image.
- Strengthening the ability to attract and retain staff or members.
- Enhancing the ability to attract and retain customers, clients or users.
- Boosting staff morale, commitment, engagement and productivity.
- More positive perceptions of the organisation by investors, donors, sponsors and the financial community.
- Improving communication and relationships with other organisations with national and local governments, with the media, with suppliers, partners, customers and with the host community.

These benefits are based on the focus of ISO 26000 on seven core subjects (Fig. 4.9), which relate to the tourism organisation's internal and external environment and relationships with external entities. Management must implement a focus on social responsibility encompassing human rights, labour practices (including employment relationships and work conditions such as health and safety), environmental concerns (such as pollution prevention, climate change mitigation, sustainable resource

Fig. 4.9 Seven core subjects of corporate social responsibility according to ISO 26000. Source: [142]

management, biodiversity protection and habitat restoration), fair operating practices (such as promoting social responsibility throughout the value chain and implementing anti-corruption policies), consumer protection (including ensuring customer health and safety and fair marketing practices) and support for local community involvement and development (including employment generation, social investment, technology development and access).

Figure 4.10 illustrates that the implementation of ISO 26000 is based on a holistic approach, encompassing the whole company and its surrounding area as well as all processes within the company. Similar is the concept of ISO 14001+, ISO 9001+, integrated management and EMAS. The implementation of ISO 26000 is also based on a systemic approach, emphasising the interrelationships between the organisation's processes, procedures and management. As in the case of ISO 14001 and ISO 9001, the implementation of ISO 26000 emphasises the key role of management and compliance with all legislative requirements that are a prerequisite for the adoption of the standard. The individual clauses of the standard and their contents are summarised in Table 4.1. The standard also contains in its annex inspiring examples of voluntary CSR initiatives. According to ISO 26000 ([145]), the standard requires organisations to respect local social, environmental, legislative, cultural, political and organisational specificities and differences in economic conditions. Figure 4.10 highlights the holistic and systemic approach with mutual interdependence of different aspects of implementing the standard. The diagram clearly demonstrates the key role of the company's management.

Fig. 4.10 Implementation of corporate social responsibility according to the individual clauses of the ISO 26000 standard. Source: Adapted from [146, 147]

Table 4.1 Content of ISO 26000

Clause title	Description of clause content
• Scope	Defines the scope of ISO 26000 and identifies certain limitations and exclusions.
• Terms and definitions	Identifies and provides the definition of key terms.
• Understanding social responsibility	Describes the important factors and conditions that have influenced the development of social responsibility and that continue to affect its nature and practice. It also describes the concept of social responsibility itself—what it means and how it applies to organisations. It includes guidance for small and medium-sized organisations on the use of this standard.
• Principles of social responsibility	Introduces and explains the principles of social responsibility.
• Recognising social responsibility and engaging stakeholders	Addresses two practices of social responsibility: an organisation's recognition of its social responsibility, and its identification of and engagement with its stakeholders.

(continued)

Table 4.1 (continued)

Clause title	Description of clause content
• Guidance on social responsibility core subjects	Explains the core subjects (see Figs. 4.9 and 4.10) and associated issues relating to social responsibility. For each core subject, information has been provided on its scope, its relationship to social responsibility, related principles and considerations, and related actions and expectations
• Guidance on integrating social responsibility throughout an organisation	Provides guidance on putting social responsibility into practice in an organisation.
Annex A Examples of voluntary initiatives and tools for social responsibility	Presents a non-exhaustive list of voluntary initiatives and tools related to social responsibility.
Annex B Abbreviated terms	Contains abbreviated terms used in ISO 26000.
Bibliography	Includes references to authoritative international instruments and ISO standards that are referenced in ISO 26000 as source material.

Source: [146]

The content of the standard is clearly shown in Fig. 4.11. It also shows the basic steps for putting each CSR topic into practice. For use in tourism, the starting point is primarily the basic practices of social responsibility, based on the tourism organisation's recognition and acceptance of its own social responsibility, the subsequent identification and involvement of tourism stakeholders, and the integration of social responsibility throughout the organisation. The understanding of social responsibility in an organisation, supported by communicating the intentions related to social responsibility to employees, suppliers, tourism participants and other tourism stakeholders, is closely related to the sustainability of tourism. The public declaration of the adoption of various voluntary instruments is very important in the tourism sector, and not only in the context of the introduction of ISO 26000.

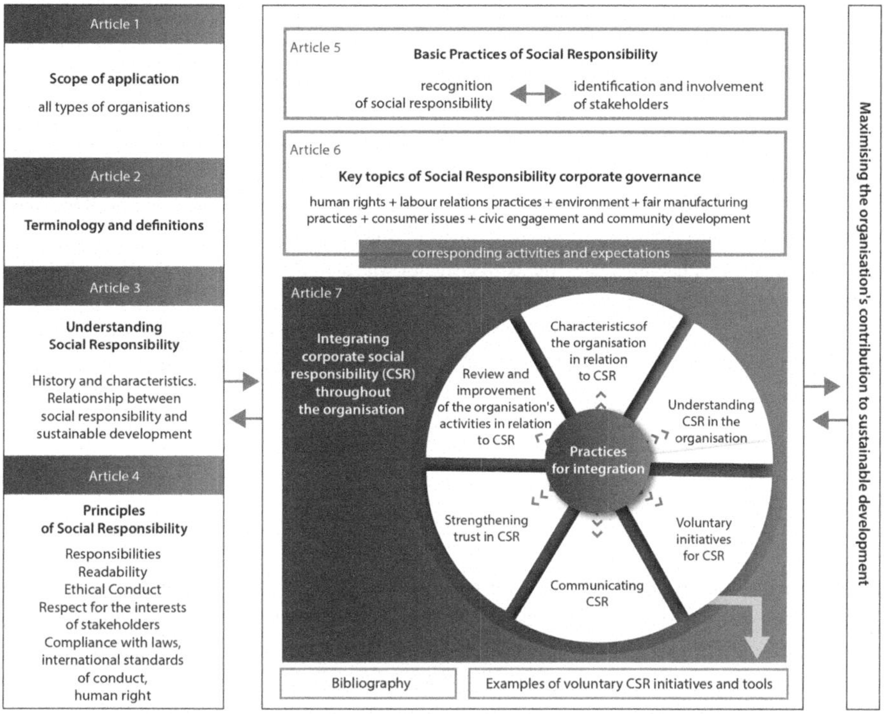

Fig. 4.11 Schematic overview of the requirements and components of ISO 26000. Source: Based on [144, 147]

4.6 Standard SA8000

According to Li [148] and Olivani and Torazzo [149], the SA8000 standard is the most important standard for CSR certification. Based on a study of its distribution in 65 countries and 66 sectors, it is expected to expand further in the coming years ([150]). Therefore, in order to make it more widely used, Llach, Marimon and Alonso-Almeida [150] suggest that it should be more flexible and respect different cultures of the world, reach more different groups of stakeholders, and encourage greater involvement of small entrepreneurs from the developing world. According to Olivani and Torazzo [149], when evaluated for all sectors of the economy, the adoption of the SA8000 standard is most widespread in Italy. However, in the tourism sector this standard is not used to a sufficient extent.

Compared to the international standard ISO 26000, the concept of the SA8000 standard is consistently based on the implementation of international agreements and the International Labour Organization (ILO) and the UN declarations on labour processes, the Universal Declaration of Human Rights of 1948, the International Covenant on Economic, Social and Cultural Rights of 1966 and the International Covenant on Civil and Political Rights of 1966 [151]. The SA8000 standard focuses on the organisation's work practices and human resource management and their practical application so that they are consistent with the principles and policies outlined in the standard for the following areas: child labour, forced or compulsory labour, health and safety, freedom of association and right to collective bargaining, discrimination, disciplinary practices, working hours, remuneration and management system [151]. Requirements include, amongst other things, the obligation to create a healthy and safe working environment, eliminate any form of employment discrimination, and ensure adequate training of employees [151].

Of particular relevance to the tourism sector are the areas of working hours, remuneration, a healthy and safe environment and child labour. This is particularly the case in hospitality, where overtime, inadequate remuneration, the use of child labour and the lack of health and safety in the workplace are practices often tolerated or even overlooked.

4.7 Fair Trade

A less formal than SA8000 but increasingly popular is fair trade [152]. This approach ensures fair trading conditions for suppliers from developing countries, whilst also ensuring respect for human and labour rights and environmental friendliness in product creation, in line with the SA8000 concept [152].

The most common work is on plantations, especially harvesting sugar cane, cacao and coffee beans. This work is usually remunerated with unacceptably low wages, takes place in extremely unsuitable conditions and is characterised by significant health and environmental risks [152]. Yet these crops provide the raw material basis for popular souvenirs in almost every destination in the world: chocolate and coffee products and various types of sweets. Typically, in developing countries where the fair trade approach is not respected, crops such as the aforementioned cacao, coffee or bananas are usually grown on plantations where workers' rights to decent and safe working conditions and to adequate remuneration for their work are often violated. Buying fair trade products and thus promoting them at home or whilst travelling are an expression of economic, social and environmental responsibility. It is also a clear signal to producers that consumers disagree with their unfair policy towards their employees and local environment. Organisations such as FLO (Fairtrade Labelling Organizations International; a certification scheme and database of certified producers), WFTO (World Fair Trade Organization) and EFTA (European Fair Trade Association) internationally guarantee compliance with fair trade principles, which include ([152]):

- A farmgate price corresponding to the cost of sustainable production and a decent living.
- Long-term trade relations.
- Decent working conditions.
- Compliance with International Labour Organisation (ILO) conventions.

- Ban on forced and child labour.
- Sustainable development of local communities.
- Controlled use of pesticides and other agricultural chemicals.
- Prohibition of the use of genetically modified crops.
- Sustainable approach to the use of natural resources.
- Democratic decision-making of local communities.
- Gender equality and equality by nations, nationalities and ethnic minorities.

As various studies show (e.g. [153]), knowledge of fair trade principles is becoming widespread in the most developed countries (unlike, e.g. Turkey, see [30]), it is an important motivating principle for customer behaviour and influencing factor of his/her experience. Customers more readily accept an increase in the price of fair trade products than an increase in taxes, and at the same time their declared purchasing behaviour matches their real behaviour [153]. Ladhari and Tchetgna [154] in their research on customers' socio-economic characteristics, confirmed the commonly expected assumption that consumption of fair trade products is higher amongst customers with higher values of equality and social justice. However, according to their findings, this consumption decreases inversely proportional to their power and social status. In this sense, their research challenges the often-held myth that fair trade is more likely to be practised by wealthier people. According to Kirezli and Kuşcu [30], greater growth in fair trade can be achieved by making fair trade products more widely available and understanding its nature, which will allow higher prices to be accepted.

In tourism, fair trade can be significantly supported by tourists' preference for fair trade products as souvenirs and gifts, as well as by their preference for catering, accommodation and other services provided by fair trade suppliers.

4.8 Environmental Management

EMS forms an important part of the implementation of corporate social responsibility. Blanco, Rey-Maquieira and Lozano [155] point to numerous examples of cost-effective EMS implementation in tourism. Its development has several directions or practical ways of implementation:

- **Cleaner production methods** (Sect. 4.8.3) primarily increase the efficiency of input resource use and reduce environmental impacts, resulting in environmental and economic benefits for the company.
- Different forms of **green management** (Sect. 4.8.2) based on a pragmatic approach to finding problematic processes and procedures using prepared lists of questions (checklists).
- **System approach**s using alternative standardised approaches with the specification of general procedures for the needs of a particular company, or systems and holistic TQM approaches with a focus on environmental issues (TQEM).

Conceptual approaches to environmental management are illustrated in Fig. 4.12.

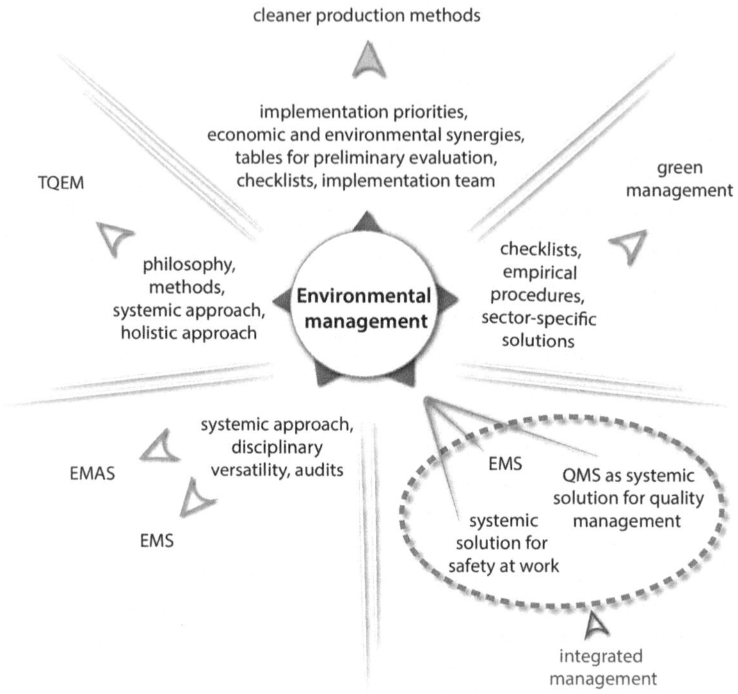

Fig. 4.12 Conceptual approaches to environmental management

4.8.1 Systemic Concept of Environmental Management

The systemic concepts of environmental management, based on standardised approaches, offer two similar concepts in terms of structure of requirements and method of implementation—EMAS (environmental management system for the EU) and EMS (based on ISO 14001). The explanation for EMS according to the international standard ISO 14001, given below, applies similarly to EMAS. EMAS also contains requirements for environmental declarations and sets out the frequency of audits.

The basic international standard for the implementation of an EMS is ISO 14001, which is followed by other standards for third-party EMS auditing, EMS implementation methodology or product life cycle. This whole system of standards is then referred to as ISO 14001+. In developing an EMS, each organisation adapts the general requirements of ISO 14001 to its needs by setting appropriate objectives, developing its environmental policy, and describing processes and procedures in the EMS environmental manual and other documentation. EMS also includes appropriate organisational structures, planning activities, responsibilities, procedures described with an emphasis on the prevention of negative environmental impacts and resources for implementation.

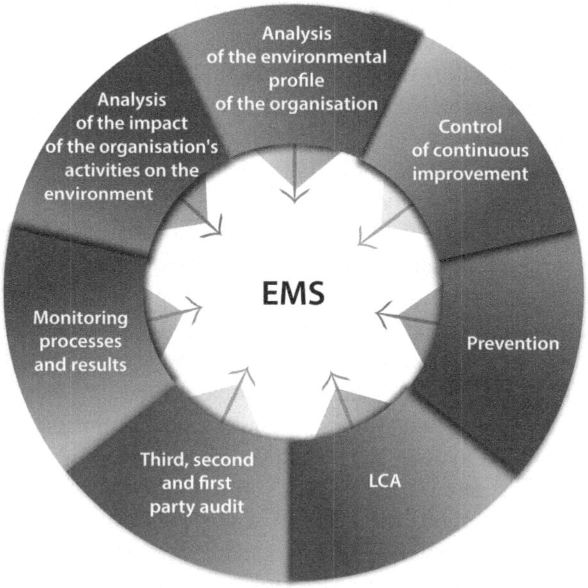

Fig. 4.13 Multifaceted feedback of operational, tactical and strategic evaluation of the EMS

EMS tools should make EMS implementation more efficient and effective. Organisations can use some of them according to their needs. Typical EMS tools include the following (Fig. 4.13):

- **Environmental audit** (internal evaluation, customer and certification body evaluation).
- **Monitoring** of processes and results achieved.
- **Analysis of the environmental impact of the organisation's activities**, including the EIA method in the case of investment and reengineering.
- **Analysis of the environmental profile of the organisation.**
- **Monitoring the continuous improvement of the organisation's environmental profile** in accordance with the requirements of ISO 14001.
- **Life cycle analysis of the product** including service, with regard to environmental impacts—LCA method.
- Emphasis on a **preemptive strategy** as part of all activities.
- **Environmental labelling of products**.
- **Management of internal and** increasingly **external** (supply chain) **resources—** education and training of employees, suppliers, maintenance of facilities and equipment etc.
- **Internal and external communication** and interaction with clients.

Studies (e.g. [156]) confirm the positive impact of EMS implementation on the productivity of the company and its internal processes. According to these authors, the results of companies that have formally certified the EMS system are better compared to companies that have implemented EMS informally.

According to the ISO 14001+ set of standards, EMS in the tourism sector is being introduced not only in the management of companies and other entities, but also in the management of areas,[10] as illustrated by Thompson, Dumont and Gaymer [157] using the example of marine protected areas. Blanco, Rey-Maquieira and Lozano (2009) highlight the importance of declaring an environmental policy in the destination to promote voluntary EMS implementation in tourism organisations. This can increase a destination's tourism sustainability and competitiveness as visitors prefer more environmentally friendly destinations (e.g. [158]) that exhibit an optimal use of resources, an increased level of conservation and higher respect towards more sustainable lifestyle of the local and Indigenous communities. The adoption of ISO 14001+ is beginning to spread amongst larger airlines, and along with green management and cleaner production methods in hotel chains. According to the ICAO report on the use of EMS in air transport ([49]), ISO 14001+ has been implemented by China Airlines, SAS, KLM, Air France and American Airlines, amongst others. Amongst hotel chains, Ibis was the first in the world to achieve ISO 14001+ certification in 2004 ([159]); other chains include Hilton Hotels & Resorts and Nordic Choice Hotels.

Cleaner production methods (Sect. 4.8.3) or green management (Sect. 4.8.2) can be the first stage of EMS implementation to address priority environmental issues. They can also help to secure funding for the implementation of an EMS system solution according to ISO 14001+ or EMAS.

In recent years, these approaches have been linked to the implementation of other system requirements for company processes, focused on quality (QMS) and occupational health and safety (ISO 45001:2018/OHSAS 18001). All this has led to the use of integrated management ([20, 160, 161]) that involves systemic approaches based on EMS, QMS and OHSAS standards. The basic standard for QMS (Quality Management System) is ISO 9001, which is linked to other standards for auditing, or for implementation methodology. The whole set of standards is referred to as ISO 9000+.

Environmental management as a systemic management concept has been developing in relation to quality management since the 1980s and was formalised in international ISO standards in the early 1990s. In 1995, its form for EU countries, EMAS, came into force, and in 2001 it was issued in a new form as Regulation No. 761/2001 of the European Parliament and of the Council ([162]). A significant trend, which is also illustrated in the diagram in Fig. 4.12, is the gradual convergence and integration of EMS with quality management (QMS) since the 1990s. This convergence includes similar approaches, basic documentation, implementation schemes and processes, management system requirements and audits of both systems. This enabled the joint implementation of both systems at the beginning of the new millennium (for details, see, e.g. [20, 160]) and the gradual implementation of integrated management, the basic scheme of which has been described by Zelenka [161] and Zelenková et al. [20]. The historical development of quality management and its eventual gradual transformation into integrated management is shown in Fig. 4.14.

[10] Protected areas, tourism destinations etc.

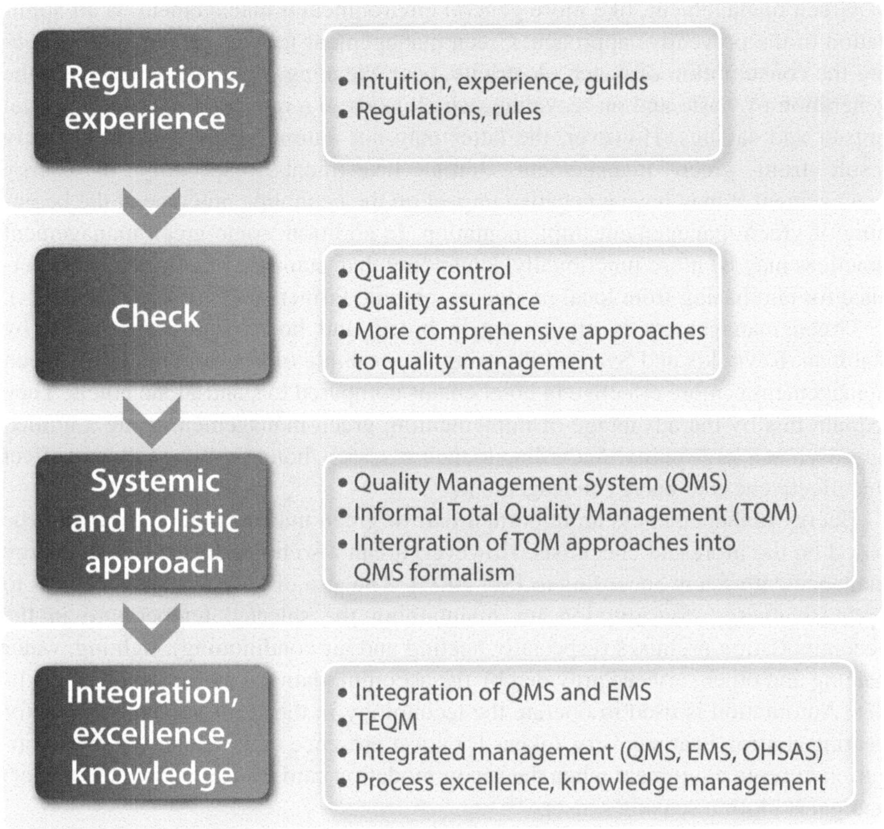

Fig. 4.14 Historical development of quality management (top to bottom) and its integration with EMS and OHSAS into integrated management. Source: [20, 160]

TQEM (Total Quality Environmental Management) is the environmental application of TQM approaches. The basis of TQEM is in line with similar principles of TQM and EMS according to ISO 14001 in organisations:

- A systemic approach, involving all employees of the organisation.
- Prevention of environmental risks and negative environmental impacts.
- Continuous improvement of the environmental profile of the organisation and the way processes are conducted in the organisation.

4.8.2 Green Management

The simplest approach to implementing environmental social responsibility is to draw on codified experience of good practice, written in the form of specific recommendations and relevant questions (for hotels, e.g. [163]). The questions are grouped into key environmental headings, which are usually accompanied by feedback checklists.

Green management, like more general environmental management, is an application of the preventive approach. Green management focuses primarily on reducing the consumption of water, electricity, heat, cleaning products, on reducing the generation of waste and on recycling, which leads to a reduction in environmental impact and savings. However, the latter may not automatically and immediately result from green management. Initial investments, especially in energy management,[11] may have a negative impact on the economic outcome at the beginning of green management implementation. In addition, some green management practices may be more functionally demanding than standard practices (e.g. preference for purchasing from local producers, organic farmers and fair trade suppliers).

Green management is often used in hotels and hotel chains. As shown by Rahman, Reynolds and Svaren [48] using the example of hotels in the USA, green management is more effective in hotel chains compared to stand-alone hotels. They explain this by the advantage of implementing green management using a unified approach in a large entity. According to their research, hotel size alone did not affect the effectiveness of energy management.

Energy management is an important part of green management and can also be based on the aforesaid checklists. However, it can also be performed as an energy management system according to ISO 50001. The aim of energy management is to optimise energy consumption for maintaining the selected temperature in the accommodation premises (especially heating and air conditioning), lighting, water heating and other services provided by the accommodation, catering or other facilities. Automation is used to operate the technology in the facility using sensors for heating control, heat recovery, tokens for guest presence in the room (typically activation of room equipment when the accommodation card is inserted into the reader) and guest motion sensing sensors.[12]

In the tourism industry, airlines such as China Airlines and Gulf Air have implemented energy management according to ISO 50001, as well as hotel chains such as NH Hoteles (the first in the world in 2011; [58]), Hilton Hotels & Resorts, and airports such as Brussels Zaventem Airport, Rome Fiumicino Airport and Antalya Airport in Turkey. Cruise ships are expected to integrate energy management as part of smart cruising, linking the use of hybrid ship propulsion with smart ship environments, incorporating the use of IoT, AI, automation, robotics and other technologies [164].

4.8.3 Cleaner Production Methods

Cleaner production methods are implemented not only for production processes but also for the provision of services. It is a preventive strategic and tactical approach that promotes a more efficient use of input resources. It also reduces the risks and

[11] Investments in alternative energy sources, low-energy appliances, control technology etc.

[12] For example, the light, toilet fan and flush can be switched on based on an infrared sensor signal.

immediate negative consequences of the organisation's activities on the health of employees and other stakeholders, as well as on the environment by analysing and taking measures against the identified risks [165, 166]. Regarding the economic, environmental and socio-cultural aspects in relation to the organisation or destination, it is a win-win strategy. More efficient use of input resources and reduction of risks and immediate negative effects of service provision on the environment results in a positive impact on the economy of the organisation. It helps reduce the cost of inputs and fees for the use of natural resources or pollution. Last but not least, it improves the quality and quantity of natural capital and the perception of the organisation or destination.

The basic practice of cleaner production is the gradual implementation of sub-projects, whose scheme is shown in Fig. 4.15, based on an assessment of priority needs. The content of the individual project stages is described in more detail by Šlesinger, Kozielová and Najmanová [166].

Due to the optimal resource consumption and high use efficiency, cleaner production methods receive a lot of attention in hotels and hotel chains, as shown by implementation methodologies ([168, 169]), case studies and good practice examples (e.g. [61, 168, 170]). The application of cleaner production methods leads to lower resource consumption and reduced environmental externalities such as solid and liquid waste, emissions. In particular, reductions in the production of greenhouse gases (mainly CO_2), in the use of electricity, fossil fuels and water, and a reduction in the amount of polluted water discharged have been observed, as illustrated for hotels by measurements in previously published case studies by Hogan and Bergin [168] and Gupta [61, 170].

In a hotel complex offering comprehensive services to its guests, cleaner production projects can focus on heating and its sub-components—the efficiency of heat source, heat recovery, the thermal insulation of the building, including the identification of thermal bridges and the most important causes of heat loss, and temperature control in the rooms.[13] The project can also focus on bio-waste in the restaurant—on prevention of its generation, on its management and its possible further use—beyond the legal framework in the country. This may also relate to relationships with the local community, e.g. local farmers, shelters, the Salvation Army and animal shelters. The hotel complex can also reduce the consumption of cold and hot water,[14] the consumption of cleaning and laundry detergents and energy for indoor and outdoor lighting.[15]

[13] Heating rooms to the optimum temperature and only when the guest is present, ventilation mode etc.

[14] For example, by installing taps mixing water and air.

[15] Optimal intensity and direction of light, use of motion sensors etc.

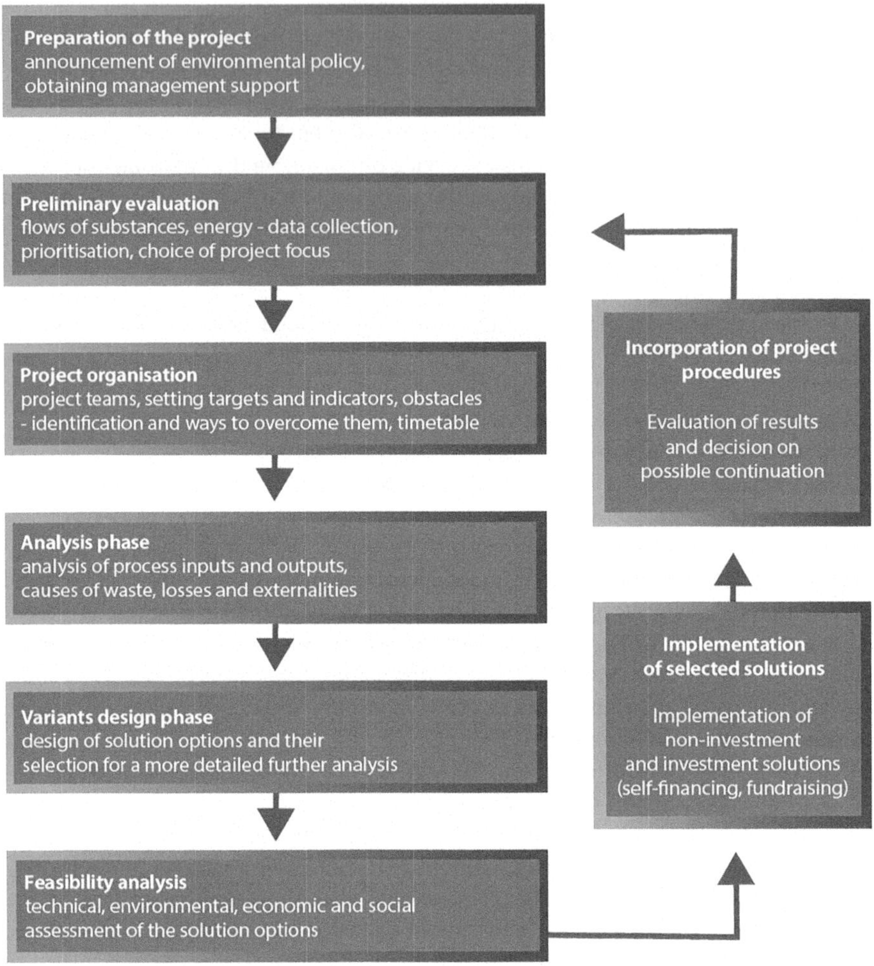

Fig. 4.15 Schematic design of a cleaner production project. Source: Adapted from [167]:19

4.9 Participatory Management

Participatory management is an attitude that contributes to sustainable development of a territory by involving all groups of stakeholders. In the case of a tourism destination, participatory management is based on systematic involvement of various destination actors. They inform each other, create common awareness and values, monitor, prepare joint projects and their implementation, generate information and knowledge for decision-making, develop alternative plans, visions, goals and objectives, and participate in the actual decision-making process. Participatory management has been developed and empirically tested in different destinations under different internal and external conditions, including the role of social capital (see

[171]). Participatory management is described in detail in, e.g. [20, 172, 173, 174, 175, 176, 177]. In destinations promoting sustainable tourism, the shared values focus strongly on social responsibility and on developing the destination's social capital (see Sect. 7.1). Furthermore, various forms of environmental management, appropriate forms of tourism and the protection of the destination's resources are promoted. According to Zelenka and Kacetl [32], participatory management is one of the basic features of quality destination/visitor management. As stated by Erkuş-Öztürk and Eraydın [178], networking at different geographical levels is increasingly important.

Hatipoglu, Alvarez and Ertuna [176] analysed the difficulties associated with the implementation and effectiveness of participatory management. They argued that the involvement of tourism actors in participatory management depends mainly on their awareness of the current state and critical aspects of tourism development, their knowledge of tourism sustainability principles and their vision of tourism planning and management in the destination. They stressed that the constraints include a narrow vision of actors, lack of strategic thinking and a focus limited to private economic gain. Amongst the external factors, the main hindrance is the lack of institutional structures supporting effective cooperation.

4.10 Incentives for Social Responsibility in Tourism

National tourist offices can support socially responsible organisations by promoting their applications for various grant programmes. They can also initiate and support CSR courses and raise awareness of businesses, local residents and visitors about CSR by means of awareness-raising campaigns or media coverage of examples of good practice.

Support for social responsibility at the global level by international organisations (e.g. UNWTO and UNESCO) and international trade fair organisers is very important. An example is the ITB Berlin tourism fair, where since 2009 the "ITB Corporate Social Responsibility Day in Tourism" showcasing the latest results and examples of practice has been organised. The fair includes lectures and events on climate change, environmentalism and sustainability in tourism. Interested parties can learn about social responsibility from information points, numerous conferences and webinars, seminars by various NGOs, and social media.

Since the 1990s, variously focused and designed certification schemes have been developed in the tourism industry as authorised procedures for reducing environmental impacts, for quality and complexity of services and for mutual comparison of services provided by different businesses and destinations. These certification schemes also serve to inform potential and real visitors and to support the development of appropriate systemic and comprehensive forms of management, and to link businesses with the local community and environment. They allow visitors to take responsibility for their contribution to the environmental impact of the service, facility and the destination in general. However, these certification schemes are fragmented and lack transparency (see e.g. [179]) and brand marketing is not very sophisticated. Only some of them are sufficiently transparent and convincing for tourism stakeholders. However, these characteristics are a fundamental condition for the success of the certification within the macro-region and its international recognition. Examples include the "Blue Flag" certification for clean and well-maintained beaches, harbours and tourist boats (see [180]) and, in Europe, the "European Flower". According to Geerts' [181] research on the adoption of environmental certification in the hotel industry, the way hotel management shares information about the objectives, content and guarantees of certification with hotel guests is crucial.

Certification schemes for ecolabelling are widely used, but as Buckley [182] points out, they are poorly coordinated, fragmented and implemented at different geographical levels. They can be viewed as codified expressions of social responsibility based on commonly declared practices such as offering organic products on organic farms and supplying organic food to organic hotels. According to Buckley [182], the impact of declared certification on visitor behaviour is poorly understood. To make certification more effective, clear ecolabel criteria need to be established with independent auditing and sanctions for breaches. This also means increasing the coverage of the tourism market by eco-certification and raising visitor awareness in terms of quality ecolabel marketing.

The use of officially recognised eco-certification is an expression of environmental social responsibility clearly recognisable from greenwashing. It can be shared by tourism actors and tourism participants. It can also enhance the competitiveness of tourism actors, as Parsa and Narapareddy [112] documented in the case of luxury resorts.

References

1. Pásková, M. (2014). *Udržitelnost cestovního ruchu [Tourism sustainability]* (p. 335). Gaudeamus Hradec Králové. ISBN 978-80-7435-329-1.
2. Young, M., Higham, J. E. S., & Reis, A. C. (2014). 'Up in the air': A conceptual critique of flying addiction. *Annals of Tourism Research, 49*, 51–64. https://doi.org/10.1016/j.annals.2014.08.003
3. Zelenka, J. (2001). Integrated management. *E+M Ekonomie a Management, 4*(1), 11–14.
4. Palmer, T., & Riera, A. (2003). Tourism and environmental taxes. With special reference to the "Balearic ecotax". *Tourism Management, 24*(6), 665–674. https://doi.org/10.1016/S0261-5177(03)00046-3
5. Park, S.-Y., & Lee, S. (2009). Financial rewards for social responsibility. A mixed picture for restaurant companies. *Cornell Hospitality Quarterly, 50*(2), 168–179. https://doi.org/10.1177/1938965509331814
6. UNWTO. (1998). Guide for local authorities on developing sustainable tourism. Madrid, p. 194 ISBN 9284402808.
7. Palhares, J. C. P., & Pezzopane, J. R. M. (2015). Water footprint accounting and scarcity indicators of conventional and organic dairy production systems. *Journal of Cleaner Production, 93*, 299–307. https://doi.org/10.1016/j.jclepro.2015.01.035

8. Pásková, M., & Dowling, R. K. (2014). The usage of local and indigenous knowledge in the management of geotourism destinations. In Conference Proceedings of 14th International Multidisciplinary Scientific Geoconference SGEM 2014, Geoconference on Ecology, Economics, Education and Legislation. Ecology and Environmental Protection. Albena, Bulgaria, Vol. II, pp. 805–812, ISBN 978-619-7105-18-6.
9. Wang, S., & Chen, J. S. (2015). The influence of place identity on perceived tourism impacts. *Annals of Tourism Research, 52*, 16–28. https://doi.org/10.1016/j.annals.2015.02.016
10. WCTE. (2022). The responsible tourist, world committee on tourism ethics, [online]. Retrieved October 4, 2022, from https://www.unwto.org/responsible-tourist
11. Wearing, S. (2001). *Volunteer tourism: Experiences that make a difference*. CAB International.
12. Garrod, B., & Fyall, A. (1998). Beyond the rhetoric of sustainable tourism? *Tourism Management, 19*(3), 199–212. https://doi.org/10.1016/S0261-5177(98)00013-2
13. Situmorang, D. B. M., & Mirzanti, I. R. (2012). Social entrepreneurship to develop ecotourism. *Procedia Economics and Finance, 4*, 398–405. https://doi.org/10.1016/S2212-5671(12)00354-1
14. Zein, K., et al. (2008). Best environmental practices for the hotel industry. Sustainable Business Associates (sba), p. 56, [online]. Retrieved January 14, 2024, from https://nrcne.org/wp-content/uploads/2019/12/SBABGEHOTELLERIEENG2008.pdf
15. Cadarso, M. A., Gómez, N., López, L. A., & Tobarra, M. Á. (2016). Calculating tourism's carbon footprint: Measuring the impact of investments. *Journal of Cleaner Production, 111*(Part B), 529–537. https://doi.org/10.1016/j.jclepro.2014.09.019
16. Tsaur, S. H., Lin, Y.-C., & Lin, J.-H. (2006). Evaluating ecotourism sustainability from the integrated perspective of resource, community and tourism. *Tourism Management, 27*(4), 640–653. https://doi.org/10.1016/j.tourman.2005.02.006
17. UNWTO. (2010). *Tourism and biodiversity. Achieving common goals towards sustainability* (p. 73). World Tourism Organization. ISBN 978-92-844-1371-3.
18. Ahn, B. Y., Lee, B., & Shafer, C. S. (2002). Operationalizing sustainability in regional tourism planning: An application of the limits of acceptable change framework. *Tourism Management, 23*(1), 1–15. https://doi.org/10.1016/S0261-5177(01)00059-0
19. Sebele, L. S. (2010). Community-based tourism ventures, benefits and challenges: Khama Rhino Sanctuary Trust, Central District, Botswana. *Tourism Management, 31*(1), 136–146. https://doi.org/10.1016/j.tourman.2009.01.005
20. Zelenka, J. (2015). *Cestovní ruch. Marketing [Tourism. Marketing]* (p. 247). Gaudeamus Hradec Králové. ISBN 978-80-7435-543-1.
21. Cohen, E., & Cohen, S. A. (2012). Current sociological theories and issues in tourism. *Annals of Tourism Research, 39*(4), 2177–2202. https://doi.org/10.1016/j.annals.2012.07.009
22. Getz, D. (1992). Tourism planning and destination life cycle. *Annals of Tourism Research, 19*(4), 752–770. https://doi.org/10.1016/0160-7383(92)90065-W
23. Li, W. (2012). Study on the relationships between corporate social responsibility and corporate international competitiveness. *Energy Procedia, 17*(Part A), 567–572. https://doi.org/10.1016/j.egypro.2012.02.137
24. Šípek, J. (2001). *Úvod do geopsychologie [An introduction to geopsychology]* (p. 163). ISV. ISBN 80-85866-70-6.
25. Vlachos, P. A., Tsamakos, A., Vrechopoulos, A. P., & Avramidis, P. K. (2009). Corporate social responsibility: Attributions, loyalty, and the mediating role of trust. *Journal of the Academy of Marketing Science, 37*(2), 170–180. https://doi.org/10.1007/s11747-008-0117-x
26. Zahra, A., & McGehee, N. G. (2013). Volunteer tourism: A host community capital perspective. *Annals of Tourism Research, 42*, 22–45. https://doi.org/10.1016/j.annals.2013.01.008
27. Pásková, M. (2014a). Geoturismus v geoparcích v ČR a ve světě [Geotourism in the Czech and world geoparks]. In D. Milošová et al. (Eds.), *Modul průvodce ekoturismu a geoturismu [Module for ecotourism and geotourism guides]*. Gaudeamus Hradec Králové. ISBN 978-80-7435-330-7, 158 s.

28. Fletcher, R., Mas, I. M., Blanco-Romero, A., & Blázquez-Salom, M. (2019). Tourism and degrowth: An emerging agenda for research and praxis. *Journal of Sustainable Tourism, 27*(12), 1745–1763. https://doi.org/10.1080/09669582.2019.1679822
29. Henderson, J. C. (2007a). Corporate social responsibility and tourism: Hotel companies in Phuket, Thailand, after the Indian Ocean tsunami. *International Journal of Hospitality Management, 26*(1), 228–239. https://doi.org/10.1016/j.ijhm.2006.02.001
30. Kerala. (2008). The Kerala declaration on responsible tourism, second international conference on responsible tourism in destinations, [online]. Retrieved February 09, 2016, from http://www.responsibletourism2008.org/keraladeclaration.php
31. Castellani, V., & Sala, S. (2010). Sustainable performance index for tourism policy development. *Tourism Management, 31*(6), 871–880. https://doi.org/10.1016/j.tourman.2009.10.001
32. Yunis. (2006). Making tourism more sustainable. Helpful instruments and practice examples in Europa. Prezentace 3. 2. 2006, Hannover, 19. s., [online]. Retrieved August 14, 2014, from http://www.ecotrans.org/docs/1_hamele_Yunis_intro_aims.pdf
33. Pásková, M., & Zelenka, J. (2018). Společensky odpovědný cestovní ruch. IDEA Servis Praha, 191 str., ISBN 978-80-85970-91-3.
34. ISO 20121:2012: Event sustainability management systems – Requirements with guidance for use. https://www.iso.org/obp/ui/#iso:std:iso:20121:ed-1:v1:en
35. ISO 20611:2018 Adventure tourism – Good practices for sustainability, requirements and recommendations.
36. Caton, K. (2012). Taking the moral turn in tourism studies. *Annals of Tourism Research, 39*(4), 1906–1928. https://doi.org/10.1016/j.annals.2012.05.021
37. Arnaudov, K., & Koseska, E. (2012). Business ethics and social responsibility in tourist organizations in terms of changing environment. *Procedia – Social and Behavioral Sciences, 44*, 387–397. https://doi.org/10.1016/j.sbspro.2012.05.043
38. Jamal, T., & Higham, J. (2021). Justice and ethics: Towards a new platform for tourism and sustainability. *Journal of Sustainable Tourism, 29*(2–3), 143–157. https://doi.org/10.1080/09669582.2020.1835933
39. Mostafanezhad, M. (2014). Volunteer tourism and the popular humanitarian gaze. *Geoforum, 54*, 111–118. https://doi.org/10.1016/j.geoforum.2014.04.004
40. Pearce, P. L., & Lee, U. (2005). Developing the travel career approach to tourist motivation. *Journal of Travel Research, 43*(3), 226–237. https://doi.org/10.1177/0047287504272020
41. Jehlička, P., Tomeš, L., & Daněk, P. (2000). *Stát, prostor, politika. [State, space, politics]* (p. 270). Přírodovědecká fakulta UK Prague.
42. Saenz-de-Miera, O., & Rosselló, J. (2014). Modeling tourism impacts on air pollution: The case study of PM10 in Mallorca. *Tourism Management, 40*, 273–281. https://doi.org/10.1016/j.tourman.2013.06.012
43. Kajala, L., et al. (2004). Measures for developing sustainability of nature tourism in protected areas. In Proceedings of the second international conference on monitoring and management of visitor flows in recreational and protected areas, June 16–20, Rovaniemi, Finland, p. 417, 236–241. ISBN 951-40-1930-X (PDF), [online]. Retrieved June 24, 2013, from http://www.metla.fi/julkaisut/workingpapers/2004/mwp002.htm
44. Torres-Delgado, A., & Palomeque, F. L. (2014). Measuring sustainable tourism at the municipal level. *Annals of Tourism Research, 49*, 122–137. https://doi.org/10.1016/j.annals.2014.09.003
45. Arjomandi, A., & Seufert, J. H. (2014). An evaluation of the world's major airlines' technical and environmental performance. *Economic Modelling, 41*, 133–144. https://doi.org/10.1016/j.econmod.2014.05.002
46. Young, A., & Kielkiewicz-Young, A. (2001). Sustainable supply network management. *Corporate Environmental Strategy, 8*(3), 260–268. https://doi.org/10.1016/S1066-7938(01)00122-1
47. Liu, J., et al. (2014). The role of social capital in encouraging residents' pro-environmental behaviors in community-based Ecotourism. *Tourism Management, 41*, 190–201. https://doi.org/10.1016/j.tourman.2013.08.016

48. Pizzitutti, F., Mena, C. F., & Walsh, S. J. (2014). Modelling tourism in the Galapagos Island: An agent-based model approach. *Journal of Artificial Societies and Social Simulation, 17*(1), 14. https://doi.org/10.18564/jasss.2389
49. IATA. (2022c). IATA – Environmental Assessment (IEnvA), [online]. Retrieved August 12, 2022, from https://www.iata.org/whatwedo/environment/Pages/environmental-assessment.aspx
50. Chen, F.-Y., Chang, Y. H., & Lin, Y. H. (2012). Customer perceptions of airline social responsibility and its effect on loyalty. *Journal of Air Transport Management, 20*, 49–51. https://doi.org/10.1016/j.jairtraman.2011.11.007
51. Higham, J., Cohen, S. A., Cavaliere, C. T., Reis, A., & Finkler, W. (2016). Climate change, tourist air travel and radical emissions reduction. *Journal of Cleaner Production, 111*(Part B), 336–347. https://doi.org/10.1016/j.jclepro.2014.10.100
52. IATA. (2022a). Enabling voluntary carbon offsetting, [online]. Retrieved August 12, 2022, from https://www.iata.org/en/programs/environment/carbon-offset/
53. Lindberg, K., Enriquez, J., & Sproule, K. (1996). Ecotourism questioned: Case studies from Belize. *Annals of Tourism Research, 23*(3), 543–562. https://doi.org/10.1016/0160-7383(95)00074-7
54. IATA. (2022b). Developing Sustainable Aviation Fuel (SAF), [online]. Retrieved August 12, 2022, from https://www.iata.org/en/programs/environment/sustainable-aviation-fuels/
55. Hahn, H., & Kagelmann, H. J. (1993). *Tourismuspsychologie und Tourismussociologie: ein Handbuch zur Tourismuswissenschaft* (p. 630). Quintessenz.
56. Mariot, P. (1983). *Geografia cestovného ruchu* (p. 249). Veda.
57. Priestley, G., & Mundet, L. (1998). The post-stagnation phase of the resort cycle. *Annals of Tourism Research, 25*(1), 85–111. https://doi.org/10.1016/S0160-7383(97)00062-5
58. denkstatt. (2016). The green hotel chain: Environmental- and energy management system implementation. [online]. Retrieved February 14, 2016, from http://denkstatt-group.com/projects/the-green-hotel-chain-environmental-and-energy-management-system-implementation/
59. Weaver, D. B. (1990). Grand Caymann Island and the resort cycle concept. *Journal of Travel Research, 29*(2), 9–15. https://doi.org/10.1177/004728759002900202
60. Rodríguez, F. J. G., & Cruz, Y. M. A. (2007). Relation between social-environmental responsibility and performance in hotel firms. *International Journal of Hospitality Management, 26*(4), 824–839. https://doi.org/10.1016/j.ijhm.2006.08.003
61. Gupta, P. K. (2014a). Resource efficient cleaner production: Case study hotel sector. Hotel Manoluck-Luang Prabang. Cleaner Production Center Lao PDR, p. 8, [online]. Retrieved January 14, 2016, from http://www.laocpc.org/index.php/publication-3/case-study
62. Bohdanowicz, P., & Martinac, I. (2007). Determinants and benchmarking of resource consumption in hotels – Case study of Hilton International and Scandic in Europe. *Energy and Buildings, 39*(1), 82–95. https://doi.org/10.1016/j.enbuild.2006.05.005
63. Chan, W. W. (2009). Environmental measures for hotels' environmental management systems: ISO 14001. *International Journal of Contemporary Hospitality Management, 21*(5), 542–560. https://doi.org/10.1108/09596110910967791
64. Charter. (1995). World Charter for Sustainable Tourism, World Conference on Sustainable Tourism, Lanzarote, Canary Islands, Spain, pp. 27–28. 4 1995, [online]. Retrieved August 04, 2022, from http://www.gdrc.org/uem/eco-tour/charter.html
65. Grimwood, B. S. R., Yudina, O., Muldoon, M., & Qiu, J. (2015). Responsibility in tourism: A discursive analysis. *Annals of Tourism Research, 50*, 22–38. https://doi.org/10.1016/j.annals.2014.10.006
66. Černec, M. (2014). Corporate social responsibility practices in international hotel chains: A web content analysis. Dissertation, p. 126, [online]. Retrieved February 09, 2016, from http://www.cek.ef.uni-lj.si/magister/Cernec1527-B.pdf
67. Chan, E. S. W., & Wong, S. C. K. (2006). Motivations for ISO 14001 in the hotel industry. *Tourism Management, 27*(3), 481–492. https://doi.org/10.1016/j.tourman.2004.10.007

68. Chan, E. S. W. (2008). Barriers to EMS in the hotel industry. *International Journal of Hospitality Management, 27*(2), 187–196. https://doi.org/10.1016/j.ijhm.2007.07.011
69. Kreisel, B. (2001). Generating income in protected areas. Issues of sustainable tourism development, EUROPARC: Loving them to death? Sustainable tourism in Europe's nature and national parks, Grafenau 2001, [online]. Retrieved July 5, 2023, from http://www.macaulay.ac.uk/workshop/bulgaria02/bettinakreisel_bulgaria02.pdf
70. Jørgensen, M. T., Hansen, A. V., Sørensen, F., Fuglsang, L., Sundbo, J., & Jensen, J. F. (2021). Collective tourism social entrepreneurship: A means for community mobilization and social transformation. *Annals of Tourism Research, 88*, 103171. https://doi.org/10.1016/j.annals.2021.103171
71. Tarí, J. J., Claver-Cortés, E., Pereira-Moliner, J., & Molina-Azorín, J. F. (2010). Levels of quality and environmental management in the hotel industry: Their joint influence on firm performance. *International Journal of Hospitality Management, 29*(3), 500–510. https://doi.org/10.1016/j.ijhm.2009.10.029
72. Chan, E. S. W., & Hawkins, R. (2012). Application of EMSs in a hotel context: A case study. *International Journal of Hospitality Management, 31*(2), 405–418. https://doi.org/10.1016/j.ijhm.2011.06.016
73. Ayala, H. (1995). Ecoresort: A 'green' masterplan for the international resort industry. *International Journal of Hospitality Management, 14*(3–4), 351–374. https://doi.org/10.1016/0278-4319(95)00043-7
74. Walker, P. A., Greiner, R., McDonald, D., & Lyne, V. (1998). The tourism futures simulator: A systems thinking approach. *Environmental Modelling & Software, 14*(1), 59–67. https://doi.org/10.1016/S1364-8152(98)00033-4
75. Carić, H. (2016). Challenges and prospects of valuation – cruise ship pollution case. *Journal of Cleaner Production, 111*(Part B), 487–498. https://doi.org/10.1016/j.jclepro.2015.01.033
76. Rastegar, R. (2022). Towards a just sustainability transition in tourism: A multispecies justice perspective. *Journal of Hospitality and Tourism Management, 52*, 113–122. https://doi.org/10.1016/j.jhtm.2022.06.008
77. CLIA. (2010). Cruise line international association, cruise market overview. Miami, FL, [online]. Retrieved August 14, 2010, from http://www.cruising.org
78. Kilipiris, F., & Zardava, S. (2012). Developing sustainable tourism in a changing environment: Issues for the tourism enterprises (Travel Agencies and Hospitality Enterprises). *Procedia – Social and Behavioral Sciences, 44*, 44–52. https://doi.org/10.1016/j.sbspro.2012.05.003
79. Dawson, J., Johnston, M. E., & Stewart, E. J. (2014). Governance of Arctic expedition cruise ships in a time of rapid environmental and economic change. *Ocean & Coastal Management, 89*, 88–99. https://doi.org/10.1016/j.ocecoaman.2013.12.005
80. Carić, H., & Mackelworth, P. (2014). Cruise tourism environmental impacts – The perspective from the Adriatic Sea. *Ocean & Coastal Management, 102*(Part A), 350–363. https://doi.org/10.1016/j.ocecoaman.2014.09.008
81. Butt, N. (2007). The impact of cruise ship generated waste on home ports and ports of call: A study of Southampton. *Marine Policy, 31*(5), 591–598. https://doi.org/10.1016/j.marpol.2007.03.002
82. Weiler, B., & Black, R. (2015). The changing face of the tour guide: One-way communicator to choreographer to co-creator of the tourist experience. *Tourism Recreation Research, 40*(3), 364–378. https://doi.org/10.1080/02508281.2015.1083742
83. Wall, G. (1983). Cycles and capacity: A contradiction in terms. *Annals of Tourism Research, 10*(2), 268–270. https://doi.org/10.1016/0160-7383(83)90030-0
84. ISO 50001:2018 Energy management systems – Requirements with guidance for use.
85. Adams, S. -A. (2014). Role of corporate social and environmental responsibility in cruise consumers' behaviour decision making. Proceedings from CTO/CREST Symposium. Third Innovators in Coastal Tourism. ICCAS July 9 – 11th 2014.
86. Joppe, M. (1996). Sustainable community tourism development revisited. *Tourism Management, 17*(7), 475–479. https://doi.org/10.1016/S0261-5177(96)00065-9

87. Lee, S., & Heo, C. Y. (2009). Corporate social responsibility and customer satisfaction among US publicly traded hotels and restaurants. *International Journal of Hospitality Management, 28*(4), 635–637. https://doi.org/10.1016/j.ijhm.2009.02.007
88. Stanners, D., & Bourdeau, P. (1995). *Europe`s environment. The Dobříš assessment* (p. 676). European Environment Agency.
89. Moghaddam, Y., Bess, C., Demirkan, H., & Spohrer, J. (2016). T-shaped: The new breed of IT professional. Cutter Consortium, [online]. Retrieved February 05, 2019, from https://www.cutter.com/article/t-shaped-new-breed-it-professional-492976
90. Campbell, J. M., & DiPietro, R. B. (2014). Sign of the times: Testing consumer response to local food signage within a casual dining restaurant. *Journal of Retailing and Consumer Services, 21*(5), 812–823. https://doi.org/10.1016/j.jretconser.2014.06.010
91. Dowd, K., & Burke, K. J. (2013). The influence of ethical values and food choice motivations on intentions to purchase sustainably sourced foods. *Appetite, 69*, 137–144. https://doi.org/10.1016/j.appet.2013.05.024
92. Phillips, P., & Moutinho, L. (2014). Critical review of strategic planning research in hospitality and tourism. *Annals of Tourism Research, 48*, 96–120. https://doi.org/10.1016/j.annals.2014.05.013
93. Rastegar, R., Higgins-Desbiolles, F., & Ruhanen, L. (2021). COVID-19 and a justice framework to guide tourism recovery. *Annals of Tourism Research, 91*, 103161. https://doi.org/10.1016/j.annals.2021.103161
94. Rezidor. (2022). The Rezidor hotel group – Code of Ethics, [online]. Retrieved April 10, 2022, from https://www.rezidor.com/sv/about-us/our-code-business-ethics/
95. Font, X., Garay, L., & Jones, S. (2016). Sustainability motivations and practices in small tourism enterprises in European protected areas. *Journal of Cleaner Production, 137*, 1439–1448. https://doi.org/10.1016/j.jclepro.2014.01.071
96. ISO 21416:2019. Recreational diving services. Requirements and guidance on environmentally sustainable practices in recreational diving.
97. Commission of the European Communities. (2003). Basic orientations for the sustainability of European tourism, COM(2003) 716 final. https://eur-lex.europa.eu/LexUriServ/LexUriServ.do?uri=COM:2003:0716:FIN:en:PDF
98. Commission of the European Communities. (2007). Agenda for a sustainable and competitive European tourism, COM(2007)621. https://www.eumonitor.eu/9353000/1/j9vvik7m1c3gyxp/vikqhlw75oyq
99. Commission of the European Communities. (2010). Communication from the Commission to the European Parliament, the Council, the European Economic and Social Committee and the Committee of the Regions – Europe, the world's No 1 tourist destination – A new political framework for tourism in Europe, COM/2010/0352 final. https://eur-lex.europa.eu/legal-content/EN/TXT/?uri=celex%3A52010DC0352
100. European Commission. (2022). Directorate-general for internal market, industry, entrepreneurship and SMEs, transition pathway for tourism, Publications Office of the European Union, [online]. Retrieved January 20, 2022, from https://data.europa.eu/doi/10.2873/344425
101. TIAC. (2022). Code of ethics and guidelines for sustainable tourism, tourism ndustry Association of Canada and National Round Table on the Environment and the Economy, [online]. Retrieved August 14, 2022, from http://nrt-trn.ca/wp-content/uploads/2011/08/sustainable-tourism.pdf
102. Morgan, N. (2012). Time for 'mindful' destination management and marketing. *Journal of Destination Marketing & Management, 1*(1–2), 8–9. https://doi.org/10.1016/j.jdmm.2012.07.003
103. Charter. (2022). The Moroccan charter of responsible tourism, [online]. Retrieved August 04, 2022, from https://www.iamorocco.com/en/portal-of-morocco/tourism/responsible-tourism/
104. Cape Town. (2016). Cape Town responsible tourism charter, [online]. Retrieved February 04, 2016, from http://www.capetown.travel/uploads/files/Res_Tourism_Charter2009.pdf
105. UNEP. (2003). Tourism and local Agenda 21. The role of local authorities in sustainable tourism, p. 59, ISBN 92-807-2267-0.

106. Cape Town. (2002). The Cape Town declaration on responsible tourism, Cape Town International Conference on Responsible Tourism in Destinations, [online]. Retrieved February 09, 2016, from http://responsibletourismpartnership.org/cape-town-declaration-on-responsible-tourism/
107. Juvan, E., & Dolnicar, S. (2014). The attitude–behaviour gap in sustainable tourism. *Annals of Tourism Research, 48*, 76–95. https://doi.org/10.1016/j.annals.2014.05.012
108. EUROPARC Federation. (2010). European Charter for sustainable tourism in protected areas, EUROPARC Federation, [online]. Retrieved August 9, 2022, from http://www.europarc.org/wp-content/uploads/2015/05/2010-European-Charter-for-Sustainable-Tourism-in-Protected-Areas.pdf
109. EGN. (2022). The EGN Charter, [online]. Retrieved August 21, 2022, from https://www.europeangeoparks.org/?page_id=357
110. EGN. (2023). Global geoparks network charter, [online]. Retrieved June 11, 2023, from https://www.europeangeoparks.org/?page_id=357
111. Council of National Geoparks. (2017). Charta národních geoparků České republiky [Charter of National Geoparks of the Czech Republic], [online]. Retrieved December 22, 2017, from http://www.geology.cz/narodnigeoparky/rada-ngcr
112. Olivani, R., & Torazzo, A. (2015). Corporate social responsibility: A survey in European environmental certified tourism facilities, [online]. Retrieved May 14, 2015, from http://forumware.wu-wien.ac.at/archiv/1348193989.pdf
113. Eurico, S. T., Oliveira, F., & Hoefnagels, H. (2014). An ethic code for European hotels: The two sides of the coin. *European Journal of Tourism, Hospitality and Recreation, 5*(3), 11–31.
114. Liu, Z. (2003). Sustainable tourism development: A critique. *Journal of Sustainable Tourism, 11*(6), 459–475. https://doi.org/10.1080/09669580308667216
115. Crang, M. (2015). Travelling ethics: Valuing harmony, habitat and heritage while consuming people and places. *Geoforum, 67*, 194–203. https://doi.org/10.1016/j.geoforum.2015.06.010
116. Pásková, M. (2022). Geopark certification as an efficient form of sustainable management of a geotourism destination. In V. Braga, A. Duarte, & C. S. Marques (Eds.), *Economics and management of geotourism. Tourism, hospitality & event management* (pp. 65–85, p. 124). Springer. https://doi.org/10.1007/978-3-030-89839-7_4
117. GGN. (2022). Code of ethics, global geoparks network, [online]. Retrieved August 4, 2022, from https://globalgeoparksnetwork.org/wp-content/uploads/2016/07/GLOBAL-GEOPARKS-NETWORK-CODE-OF-ETHICS-final.pdf
118. ARTN. (2018). Member's code of ethics – Australian regional tourism, [online]. Retrieved August 10, 2022, from https://www.tourism.australia.com/content/dam/assets/document/1/6/x/6/o/2002560.pdf
119. Sun, Y. Y. (2014). A framework to account for the tourism carbon footprint at island destinations. *Tourism Management, 45*, 16–27. https://doi.org/10.1016/j.tourman.2014.03.015
120. Talwar, S., Kaur, P., Nunkoo, R., & Dhir, A. (2022). Digitalization and sustainability: Virtual reality tourism in a post pandemic world. *Journal of Sustainable Tourism, 31*(11), 2564–2591. https://doi.org/10.1080/09669582.2022.2029870
121. ANW. (2022). Australia's North West (ANW) Code of Conduct and Ethics, [online]. Retrieved August 10, 2022, from https://www.australiasnorthwest.com/sites/nw/files/2018-05/ANW%20Code%20of%20Ethics.pdf
122. Tsai, W. H., Chang, Y.-C., Lin, S.-J., Chen, H.-C., & Chu, P.-Y. (2014). A green approach to the weight reduction of aircraft cabins. *Journal of Air Transport Management, 40*, 65–77. https://doi.org/10.1016/j.jairtraman.2014.06.004
123. Asistur. (2016). Code of Conduct for tourists in Guatemala – Hotel Princess, [online]. Retrieved April 10, 2016, from http://www.hotelesprincess.com/en/wp-content/uploads/2014/05/code-conduct-princess.pdf
124. Accor. (2022b). AccorHotels ethics and corporate social responsibility charter, [online]. Retrieved August 10, 2022, from https://group.accor.com/-/media/Corporate/Talent/PDF-for-pages/Charte-ethique-et-RSE/ACCOR_Ethics_and_Corporate_Social_Responsibility_Charter_20_ENG.pdf

125. Ramdas, M., & Mohamed, B. (2014). Impacts of tourism on environmental attributes, environmental literacy and willingness to pay: A conceptual and theoretical review. *Procedia – Social and Behavioral Sciences, 144*, 378–391. https://doi.org/10.1016/j.sbspro.2014.07.307
126. Mazurkiewicz, P., & Grenna, L. (2003). Corporate social responsibility and multi-stakeholder dialogue. Towards environmental behavioral change. World Bank, p. 20, [online]. Retrieved January 22, 2024, from https://policycommons.net/artifacts/1442709/corporate-social-responsibility-and-multi-stakeholder-dialogue/2073571/
127. Ladhari, R., & Tchetgna, N. M. (2015). The influence of personal values on fair trade consumption. *Journal of Cleaner Production, 87*, 469–477. https://doi.org/10.1016/j.jclepro.2014.10.068
128. ECPAT. (2000). The code of conduct for the protection of children from sexual exploitation in travel and tourism, [online]. Retrieved June 22, 2017, from https://www.unicef.org/lac/code_of_conduct.pdf
129. ICAO. (2022). Report on Environmental Management System (EMS) practices in the aviation sector. ICAO, p. 62. ISBN 978-92-9231-834-5, [online]. Retrieved August 12, 2022, from http://www.icao.int/environmental-protection/Documents/Publications/Doc%209968%201st%20Edition%20alltext%20en%20FINAL.pdf
130. Williams, P. W., & Ponsford, I. F. (2009). Confronting tourism's environmental paradox: Transitioning for sustainable tourism. *Futures, 41*(6), 396–404. https://doi.org/10.1016/j.futures.2008.11.019
131. Hunter, C., & Shaw, J. (2007). The ecological footprint as a key indicator of sustainable tourism. *Tourism Management, 28*(1), 46–57. https://doi.org/10.1016/j.tourman.2005.07.016
132. CityTrekkingGuide. (2022). Code of Conduct for responsible tourist, Chile, [online]. Retrieved August 22, 2022, from https://citytrekkingguide.com/wp-content/uploads/2020/09/Code-of-Conduct-for-a-responsible-tourist-%E2%80%93-Chile-%E2%80%93-CityTrekkingGuide-_-TAILORMADE-TOURS-IN-CHILE.pdf
133. TIES. (2015a). Definition of ecotourism from 1990, The International Ecotourism Society, [online]. Retrieved August 21, 2015, from https://www.ecotourism.org/book/ecotourism-definition
134. Whitla, P., Walters, P. G. P., & Davies, H. (2007). Global strategies in the international hotel industry. *International Journal of Hospitality Management, 26*(4), 777–792. https://doi.org/10.1016/j.ijhm.2006.08.001
135. Wells, V. K., et al. (2015). Heritage tourism, CSR and the role of employee environmental behaviour. *Tourism Management, 48*, 399–413. https://doi.org/10.1016/j.tourman.2014.12.015
136. ICNF. (2009). Code of conduct for nature tourism companies, [online]. Retrieved August 22, 2022, from https://www.icnf.pt/api/file/doc/d0fda142466ed5b2
137. Luo, W., Tang, P., Jiang, L., & Su, M. M. (2020). Influencing mechanism of tourist social responsibility awareness on environmentally responsible behavior. *Journal of Cleaner Production, 271*, 122565. https://doi.org/10.1016/j.jclepro.2020.122565
138. Šlesinger, J., Kozielová, Z., & Najmanová, K. (2007). Čistší produkce. Příručka pro podniky a veřejnou správu [Cleaner Production. A guide for business and public administration]. CENIA, Prague, p. 100, [online]. Retrieved January 14, 2024, from https://invenio.nusl.cz/record/361669/files/nusl-361669_1.pdf
139. ISO 13810:2022: Tourism and related services. Visits to industrial, natural, cultural and historical sites. Requirements and recommendations.
140. ISO 13009:2015: Tourism and related services. Requirements and recommendations for beach operation.
141. ISO/TS 13811:2015: Tourism and related services — Guidelines on developing environmental specifications for accommodation establishments.
142. ISO 17680:2015: Tourism and related services. Thalassotherapy, Service requirements.
143. ISO 17679:2016: Tourism and related services. Wellness spa, Service requirements.
144. ISO 18065:2015: Tourism and related services. Tourist services for public use provided by Natural Protected Areas Authorities. Requirements.

145. Discovering ISO 26000. p. 20, [online]. Retrieved August 14, 2022, from https://www.iso.org/files/live/sites/isoorg/files/store/en/PUB100258.pdf
146. ISO 26000 and the international integrated reporting <IR> framework briefing summary, ISBN 978-92-67-10635-9, p. 12, [online]. Retrieved August 14, 2022, from https://www.iso.org/files/live/sites/isoorg/files/store/en/PUB100402.pdf.
147. Schematic overview of ISO 26000. (n.d.). [online]. Retrieved August 14, 2022, from https://www.iso.org/publication/PUB100260.html
148. Lee, K. H., Packer, J., & Scott, N. (2015). Travel lifestyle preferences and destination activity choices of slow food members and non-members. *Tourism Management, 46*, 1–10. https://doi.org/10.1016/j.tourman.2014.05.008
149. Naess, A. (2009). *Ecology, community and lifestyle. Outline of ecosophy* (p. 223). Cambridge University Press. https://doi.org/10.1017/CBO9780511525599
150. Li, T.-T., Wang, K., Sueyoshi, T., & Wang, D. D. (2021). ESG: Research progress and future prospects. *Sustainability, 13*(21), 11663. https://doi.org/10.3390/su132111663
151. Singh, A. K., Zhang, Y., & Anu. (2023). Understanding the evolution of environment, social and governance research: Novel implications from bibliometric and network analysis. *Evaluation Review, 47*(2), 350–386. https://doi.org/10.1177/0193841X221121244
152. Fair Trade. (2022). Fairtrade international, [online]. Retrieved August 14, 2022, from https://www.fairtrade.net/
153. Campbell, C. L., Heinrich, D., & Schoenmüller, V. (2015). Consumers' reaction to fair trade motivated price increases. *Journal of Retailing and Consumer Services, 24*, 79–84. https://doi.org/10.1016/j.jretconser.2015.02.005
154. Klein, R. A. (2011). Responsible cruise tourism: Issues of cruise tourism and sustainability. *Journal of Hospitality and Tourism Management, 18*(1), 107–116. https://doi.org/10.1375/jhtm.18.1.107
155. Blanco, E., Rey-Maquieira, J., & Lozano, J. (2009). Economic incentives for tourism firms to undertake voluntary environmental management. *Tourism Management, 30*(1), 112–122. https://doi.org/10.1016/j.tourman.2008.04.007
156. Mayer, R., Ryley, T., & Gillingwater, D. (2012). Passenger perceptions of the green image associated with airlines. *Journal of Transport Geography, 22*, 179–186. https://doi.org/10.1016/j.jtrangeo.2012.01.007
157. Sucheran, R. (2021). Global impacts and trends of the COVID-19 pandemic on the cruise sector: A focus on South Africa. *African Journal of Hospitality, Tourism and Leisure., 10*(1), 22–39. https://doi.org/10.46222/ajhtl.19770720-84
158. Nicolae, J. C., & Sabina, D. B. M. (2012). The ICT implication on CSR in the tourism of emerging markets. *Procedia Economics and Finance, 3*, 702–709. https://doi.org/10.1016/S2212-5671(12)00217-1
159. Accor. (2022a). AccorHotels – brands – Ibis, [online]. Retrieved August 4, 2022, from http://www.accorhotels-group.com/en/brands/brand-portfolio/ibis.html
160. Domingues, J. P. T., Sampaio, P., & Arezes, P. M. (2015). Analysis of integrated management systems from various perspectives. *Total Quality Management & Business Excellence, 26*(11–12), 1311–1334. https://doi.org/10.1080/14783363.2014.931064
161. Xu, J., Yao, L., & Mo, L. (2011). Simulation of low-carbon tourism in world natural and cultural heritage areas: An application to Shizhong District of Leshan City in China. *Energy Policy, 39*(7), 4298–4307. https://doi.org/10.1016/j.enpol.2011.04.047
162. European Parliament. (2001). Regulation (EC) No 761/2001 of the European Parliament and of the Council of 19 March 2001 allowing voluntary participation by organisations in a Community eco-management and audit scheme (EMAS), [online]. Retrieved June 14, 2017, from https://eur-lex.europa.eu/legal-content/EN/TXT/?uri=CELEX:32001R0761
163. Xing, Y., & Dangerfield, B. (2010). Modelling the sustainability of mass tourism in island tourist economies. *Journal of the Operational Research Society, 62*(2011), 1742–1752. https://doi.org/10.1057/jors.2010.77
164. Tribe, J. (2009). *Philosophical issues in tourism* (p. 272). Channel View Publications. ISBN 9781845412494.

165. Hens, L., Block, C., Cabello-Eras, J. J., Sagastume-Gutierez, A., Garcia-Lorenzo, D., Chamorro, C., Herrera Mendoza, K., Haeseldonckx, D., & Vandecasteele, C. (2018). On the evolution of "Cleaner Production" as a concept and a practice. *Journal of Cleaner Production, 172*, 3323–3333. https://doi.org/10.1016/j.jclepro.2017.11.082
166. Sinclair-Maragh, G., & Gursoy, D. (2015). Imperialism and tourism: The case of developing island countries. *Annals of Tourism Research, 50*, 143–158. https://doi.org/10.1016/j.annals.2014.12.001
167. Singal, M. (2014). Corporate social responsibility in the hospitality and tourism industry: Do family control and financial condition matter? *International Journal of Hospitality Management, 36*, 81–89. https://doi.org/10.1016/j.ijhm.2013.08.002
168. Hogan, J., & Bergin, M. (2007). Development of a cleaner production programme for the Irish hotel industry – Greening Irish Hotels. Environmental Protection Agency, ISBN 1-84095-225-3, [online]. Retrieved January 14, 2016, from http://ctc-cork.ie/File/greening_irish_hotels.pdf
169. Noronha, R. (1979). *Social and cultural dimensions of tourism* (p. 63). World Bank.
170. Gupta, P. K. (2014b). Resource Efficient cleaner production: Case study hotel sector. Hotel Da Le Paix-Luang Prabang. Cleaner Production Center Lao PDR, p. 8, [online]. Retrieved January 14, 2016, from http://www.laocpc.org/index.php/publication-3/case-study
171. Zelenka, J., et al. (2014). Aplikace umělé inteligence a kognitivní vědy v udržitelnosti cestovního ruchu [Application of artificial intelligence and cognitive science in tourism sustainability]. *Gaudeamus Hradec Králové*, 219. ISBN 978-80-7435-354-3.
172. Beritelli, P. (2011). Cooperation among prominent actors in a tourist destination. *Annals of Tourism Research, 38*(2), 607–629. https://doi.org/10.1016/j.annals.2010.11.015
173. CBD and UNEP. (2007a). Managing tourism & biodiversity. User's manual on the CBD guidelines on biodiversity and tourism development. CBD a UNEP, ISBN 92-9225-069-8, p. 125, [online]. Retrieved August 19, 2022, from http://tourism.cbd.int/manual.shtml
174. COMMITTEE. (2002). Public participation in protected area management best practice. Parks and Wildlife Commission of the Northern Territory, [online]. Retrieved August 9, 2022, from https://www.environment.gov.au/system/files/resources/45209a20-1ab1-40d1-aa26-cf4838948782/files/public-participation.doc
175. Eagles, P. J. F., McCool, S. F., & Haynes, C. D. (2002). Sustainable tourism in protected areas: Guidelines for planning and management. UNEP/IUCN (World Commission on Protected Areas)/WTO, Best Practice Protected Area Guidelines Series No. 8, p. 183, ISBN 2-8317-0648-3, [online]. Retrieved October 8, 2011, from www.uneptie.org/pc/tourism/library/sust_prot_areas.htm, www.iucn.org/dbtwwpd/-edocs/PAG-008.pdf
176. Hatipoglu, B., Alvarez, M. D., & Ertuna, B. (2016). Barriers to stakeholder involvement in the planning of sustainable tourism: The case of the Thrace region in Turkey. *Journal of Cleaner Production, 111*(Part B), 306–317. https://doi.org/10.1016/j.jclepro.2014.11.059
177. Václav Havel Airport Prague. (2023). Prague airport's development quest: Sustainable and Gree, [online]. Retrieved May 26, 2023, from https://www.prg.aero/en/letiste-praha-se-rozviji-udrzitelne-do-zelenych-barev
178. Erkuş-Öztürk, H., & ERAYDIN, A. (2010). Environmental governance for sustainable tourism development: Collaborative networks and organisation building in the Antalya tourism region. *Tourism Management, 31*(1), 113–124. https://doi.org/10.1016/j.tourman.2009.01.002
179. Font, X. (2002). Environmental certification in tourism and hospitality: Progress, process and prospects. *Tourism Management, 23*(3), 197–205. https://doi.org/10.1016/S0261-5177(01)00084-X
180. Blue Flag. (2022). Blue flag. [online]. Retrieved August 14, 2022, from http://www.blueflag.org/
181. Geerts, W. (2014). Environmental certification schemes: Hotel managers' views and perceptions. *International Journal of Hospitality Management, 39*, 87–96. https://doi.org/10.1016/j.ijhm.2014.02.007
182. Buckley, R. (2002). Tourism ecolabels. *Annals of Tourism Research, 29*(1), 183–208. https://doi.org/10.1016/S0160-7383(01)00035-4

Chapter 5
Case Studies of Social Responsibility in Tourism

5.1 Social Responsibility in the Hospitality Sector

As evidenced by Bodhanwala and Bodhanwala [1], the hotel industry stands out for the highest use of ESG approaches in the implementation of social responsibility amongst tourism sectors. ESG-Compliant Standards for Hotel Certification are being created, as described by Bae [2] for the Republic of Korea.

As reported by Lee and Heo [3] for hotels and restaurants, the consistent application of corporate responsibility increases guest satisfaction and the value of the business. De Grosbois [4] analysed 150 hotel companies and pointed out the fundamental difference between CSR proclamation and consistent description and disclosure of the actual situation. According to him, only a small number of hotel companies that claim to be committed to social responsibility describe their activities to realise it, and even fewer describe the actual results of these activities. As examples of good practice can serve IHG [5] and Four Seasons [6], which clearly declare both their awareness of the environmental impacts of their activities and their intention to actively approach these impacts together with their suppliers, guests and other relevant stakeholders. They also share the results achieved. Martínez and del Bosque [7] concluded that social responsibility significantly influences the evaluation of the product provided, customer satisfaction, trust in the hotel and customer identification with the hotel. Levy and Park [8] consider that the greatest benefits of adopting social responsibility in American hotels are economic savings and improved customer perception of the hotel brand. Improved customer perception of the hotel brand confirmed Akbari et al. [9] for UK hotels. According to Levy and Park [8], hotels focus mainly on energy conservation, waste management and water management.

Kucukusta, Mak and Chan [10] found in the example of guests of Hong Kong four- and five-star hotels which components and conditions of the implementation of social responsibility are perceived and what importance is attached to them.

M. Pásková, J. Zelenka, *Social Responsibility in Tourism*, Tourism on the Verge,
https://doi.org/10.1007/978-3-031-61609-9_5

According to them, the declared vision and mission and environmental measures had the greatest impact on length of stay, willingness to pay and perceptions of service quality and brand image. Lee et al. [11] examined guest perceptions of hotels adopting green management.[1] They concluded that quality attributes have a greater influence on guests than value attributes and that "selfish altruism" and feeling good about the purchase are important to guests.

At present, however, the influence of hotels' declared social responsibility on the preferences of guests cannot be overestimated, as only a relatively small part of guests is influenced by declared CSR ([13, 14]). It is also important to use marketing effectively, including an emphasis on green hotel management, which increases brand loyalty ([15]).

Social responsibility for hotels and hotel chains is illustrated by the mind map in Fig. 5.1.

Relatively little research has been done on the impact of social responsibility on hotel employees, especially on whether and how they perceive it. Research on hotels in Hong Kong has shown that hotel employees perceive most of the components of practised social responsibility, which is a condition for the influence of social responsibility on employees [16]. Aminudin [17], on the other hand, based on an analysis of Malaysian green hotels,[2] emphasises that the positive impact of social responsibility (reported in, e.g. [18]) on hotel employees should not be overestimated. Aminudin [17] suggests three key areas of influence on employees: **emotional**, including employee identification with the hotel; **reducing vagueness**, where the appropriate approach is to include employees in the decision-making process and **the creative use of different approaches to applying social responsibility to intrinsic motivation** (e.g. [19]).

Benavides-Velasco, Quintana-García and Marchante-Lara [20], based on a comparative study of Spanish hotels, showed that TQM and social responsibility increase the competitiveness of hotels and, at the same time, that TQM can promote the development of social responsibility. This is in line with the general findings on the positive impact of all three core dimensions of social responsibility ([21]) on firms' competitiveness and with the findings on the positive impact of both EMS and QMS on hotel performance ([22]). In this context, Seo, Moon and Lee [23] provide an interesting comparison with airlines. Whilst for "conventional" airlines, as for accommodation facilities, there is a positive synergy of perceived service quality and social responsibility levels with performance and competitiveness, for low-cost airlines the correlation of perceived service quality and social responsibility levels with performance is negative. This is due to the fact that this practice undermines the basis of competitiveness of low-cost airlines, which is their low cost per unit of performance.

[1] For its systematic support on the Green Hotelier website, see [12].

[2] In his view, hotels implementing green management and supporting the travel of people with disabilities.

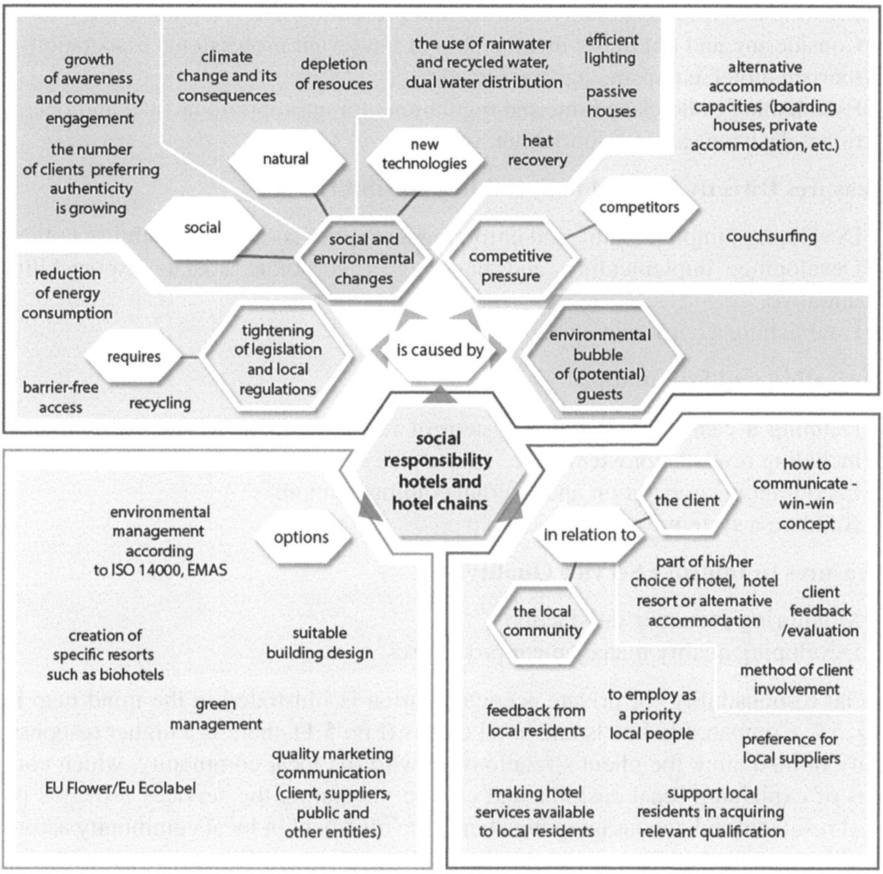

Fig. 5.1 Social responsibility—hotels and hotel chains. Source: Inspired by Kaur et al. (2022)

In order to increase the level and effectiveness of social responsibility in accommodation facilities, Matev and Assenova [24] recommend that tourism actors use the following measures in line with their level of management development and social responsibility:

Measures for the development of human resources

- Establishing a human resource development department and related policies and procedures.
- Continuous staff training, including professional development training and exchange programmes with other hotels and restaurants of the company.
- Team building.
- Survey of needs, profiles and monitoring of employee satisfaction.
- A career plan for each position.

Organisational Measures

- Creating a corporate strategy, sharing and presenting it.
- Considering and obtaining membership of a relevant professional association—tourism, hotel, campsite, agritourism etc.
- Establishing policies and internal regulations for appropriate tactical and operational decisions and for marketing.

Measures Directly Related to Social Responsibility Policy

- Developing, implementing and enforcing a supplier social responsibility policy.
- Developing, implementing and ensuring a corporate social responsibility initiative.
- Establishing a corporate code of conduct.

Motivation and Stimulation

- Defining a clear and objective system of rewards for individual performance, including rewards for excellence.
- Encouraging cooperation and internal communication.
- Building a system for evaluation.

Measures to Improve Service Quality

- Monitoring customer satisfaction.
- Developing quality management procedures.

Social responsibility for private accommodation is illustrated in the mind map in Fig. 5.2. Compared to hotels and hotel chains (Fig. 5.1), there is a higher responsibility in mediating the client's relationship with the local community, which consists of explaining local customs and culture, mediating the services provided by local residents and encouraging the client's involvement in local community activities like visiting families, social entertainment and sports. Private accommodation also offers many opportunities for social responsibility, especially by personal contact between the guest and the host, often associated with getting to know the undistorted reality of local life, as well as by providing locally specific accommodation and service built on local traditions, eco-building and strengthening the relationship to local identity by appropriate locally specific accommodation facilities.

Declared social responsibility in catering, applied mainly to large restaurants and fast food chains, is in line with their sophisticated marketing and usually includes gastronomy in addition to the general components of social responsibility ([25]). The gastronomic component of social responsibility has emerged in recent years in the context of growing knowledge about the effect of diet on human health. This new explicit part of social responsibility expresses responsibility for the health of customers as part of the responsibility towards society. It is also based on increasing legislative requirements, in particular, for consumer and environmental protections. The practice of corporate social responsibility (Fig. 5.3) is important for catering facilities and increases the share value of restaurants by reducing systemic risk ([26]). Responsibility activities in catering are summarised into three dimensions: social, environmental and gastronomic (Fig. 5.3). For catering facilities, one of the

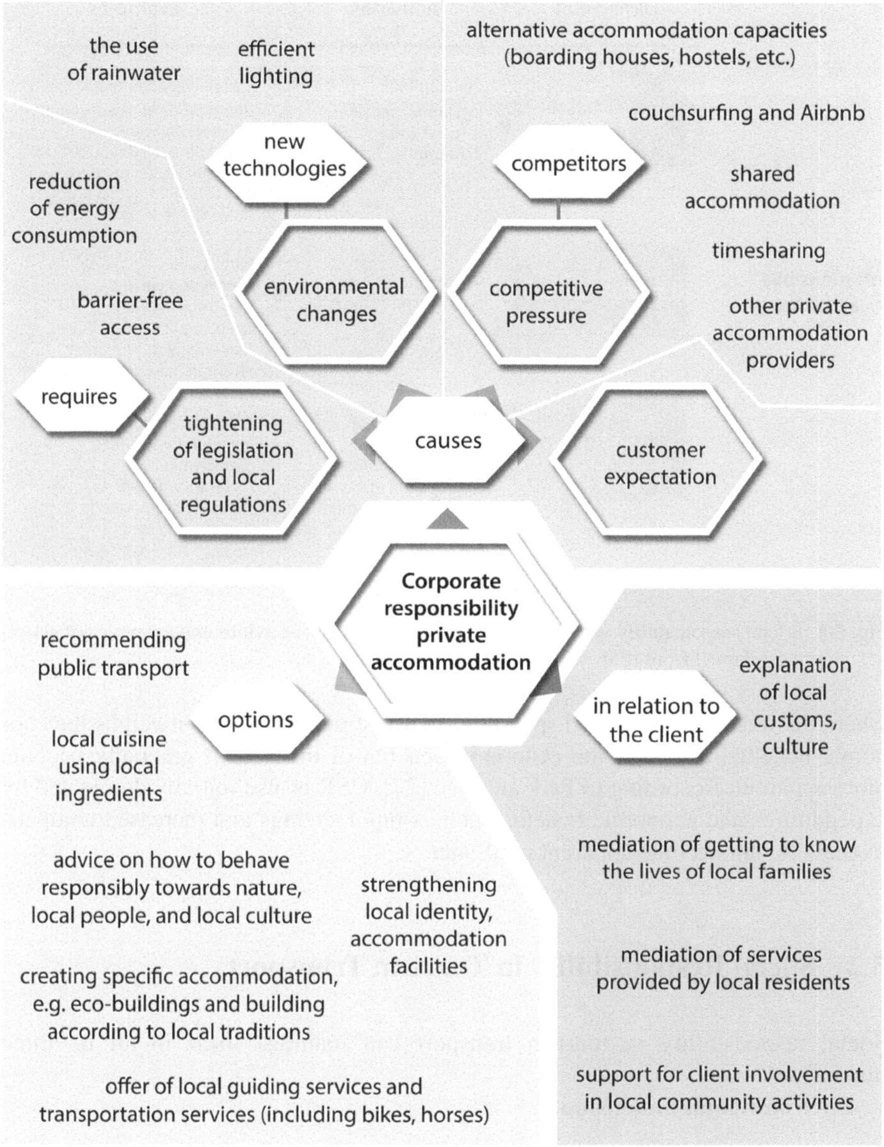

Fig. 5.2 Mind map of social responsibility for private accommodation

most important findings is the impact that the implementation of social responsibility has on their economic performance. The impact on return on equity and on total return to shareholders was investigated by Park and Lee [27]. Whilst their research found that total return to shareholders is not affected and only becomes apparent in the long run, a more significant effect was found on return on equity, in the form of a U-shaped curve. This curve describes a typical economic scenario for investing,

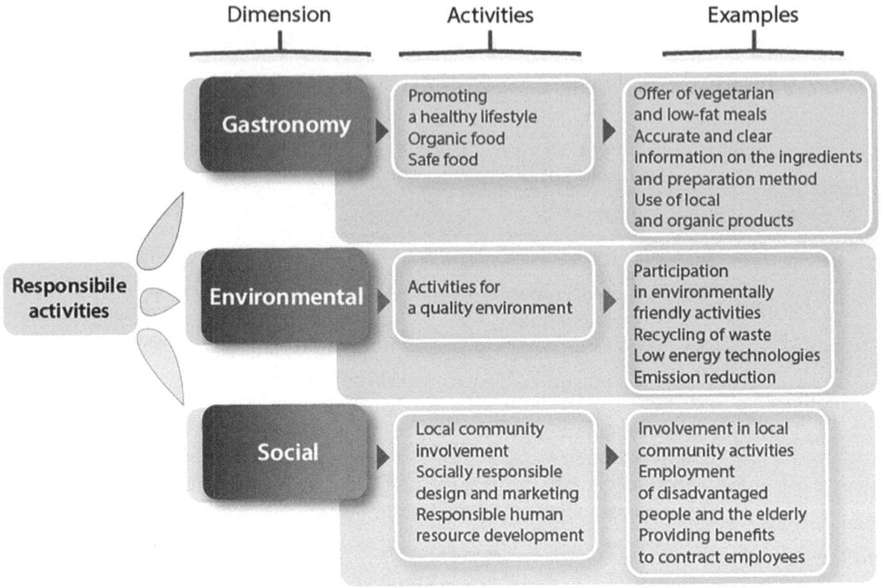

Fig. 5.3 Social responsibility in catering—dimensions, types of activities and examples of activities. Source: Adapted from [25]

where the impact of increased spending on investment initially outweighs the economic benefits, and later the economic benefits of investment gradually become more apparent. According to Park and Lee [27], CSR is also initially dominated by expenditures and economic benefits such as input savings and increased competitiveness do not become apparent until later.

5.2 Social Responsibility in Tourism Transport

Social responsibility in tourism transport can manifest itself in all its three dimensions.

Environmental dimension

- Application of environmental management, which includes modernisation of transport means and transport terminals, quality traffic management including optimisation of flight routes and appropriate planning and construction method of transport infrastructure featuring noise barriers and earth mounds, routing and tunnels, leads to a drop in air emissions, noise and light pollution, a reduction in

the impact of landscape fragmentation by linear constructions. This last-mentioned impact can be reduced by construction of migration corridors for animals under roads and railways. Collisions between vehicles and animals can be reduced due to the construction of fences around motorways.
- Promoting and encouraging more environmentally friendly forms of transport like public transport, especially train transport, and self-propelled transport by building cycle paths and pedestrian zones.
- Direct support for programmes for the protection and development of the environment, ecosystems and the biosphere including protection of tropical rainforests, various species of flora and fauna, as described by, e.g. Singapore Airlines [28], support for environmental research such as carbon and energy management (e.g. [29]), and support for environmental programmes aimed at reducing greenhouse gas emissions.

Economic dimension

- Relocation and customer-supported fundraising for environmental offset programmes. For example, IATA's Greenhouse Gas Offset Programme (see [30]). These funds can be used for such desirable activities as planting new forests.

Socio-cultural dimension

- Carriers can develop various charity programmes like direct support for education or healthcare in poor communities, support cultural development, provide scholarships for talented students, promote healthy lifestyles and sports (see, e.g. [28]). They can also prioritise employing local people, prefer local companies as their suppliers and offer locally sourced and organic food in their means of transport.

The way in which social responsibility is presented is rapidly evolving, as illustrated by changes to the concept of social responsibility on the easyJet, plc website (Fig. 5.4). In 2015, it demonstrated a very professional approach to social responsibility by implementing both preventive and offsetting activities ([31]). The former focused on reducing the environmental impact by using modern aircraft with lower emissions, fuel consumption, and less noise. The latter were dedicated to cooperation with UNICEF providing vaccinations in Africa. Preventive measures for their own employees, such as incentives to cycle to work, were inspiring. In 2022, the focus was on achieving lower CO_2 emissions in the context of the current climate change debate ([32]).

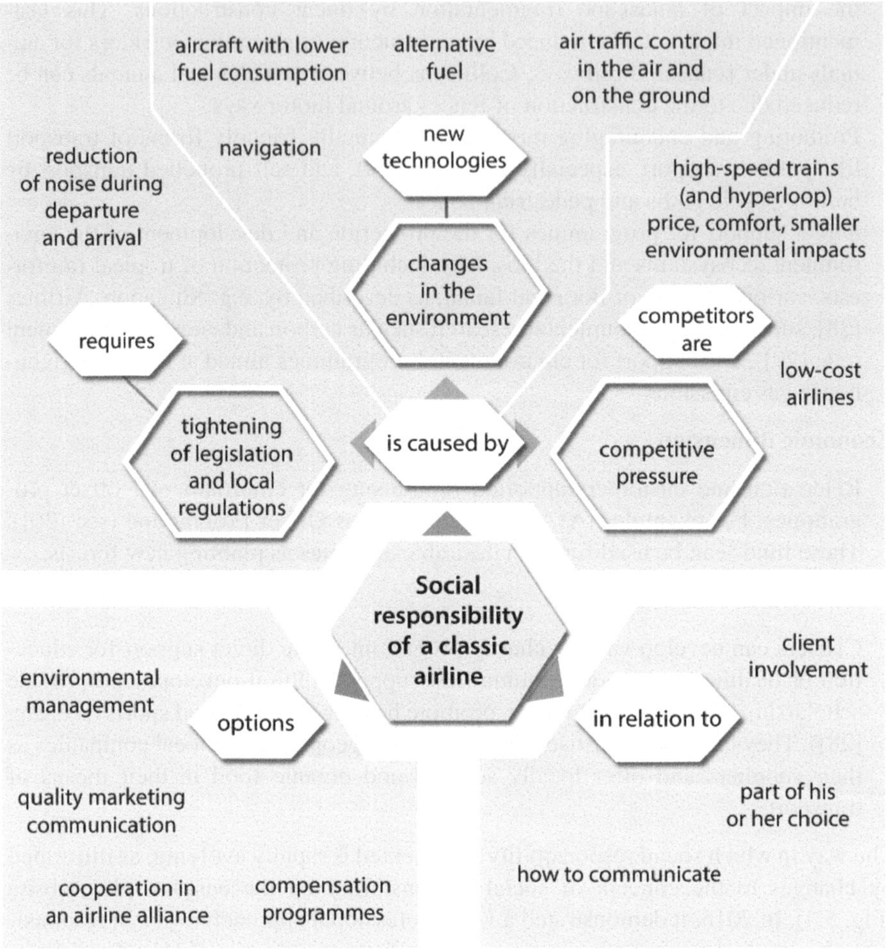

Fig. 5.4 Mind map of the social responsibility of a classic airline company

5.3 Social Responsibility of Destination Management

A destination is the basic territorial functional unit of tourism. Within it, destination management companies and other destination actors operate, and together they form destination management in its broader sense. In the context of social responsibility, a destination must be seen as an open system consisting of destination actors, primary destination resources like tourist attractions, local pretouristic infrastructure, local nature and human resources and secondary destination resources like tourism infrastructure. All these destination elements are linked by processes and relationships that ensure the internal coherence of the destination as a region. Destinations develop over time and are subject to external influences, including the political, security and economic situation of the country and the level of its

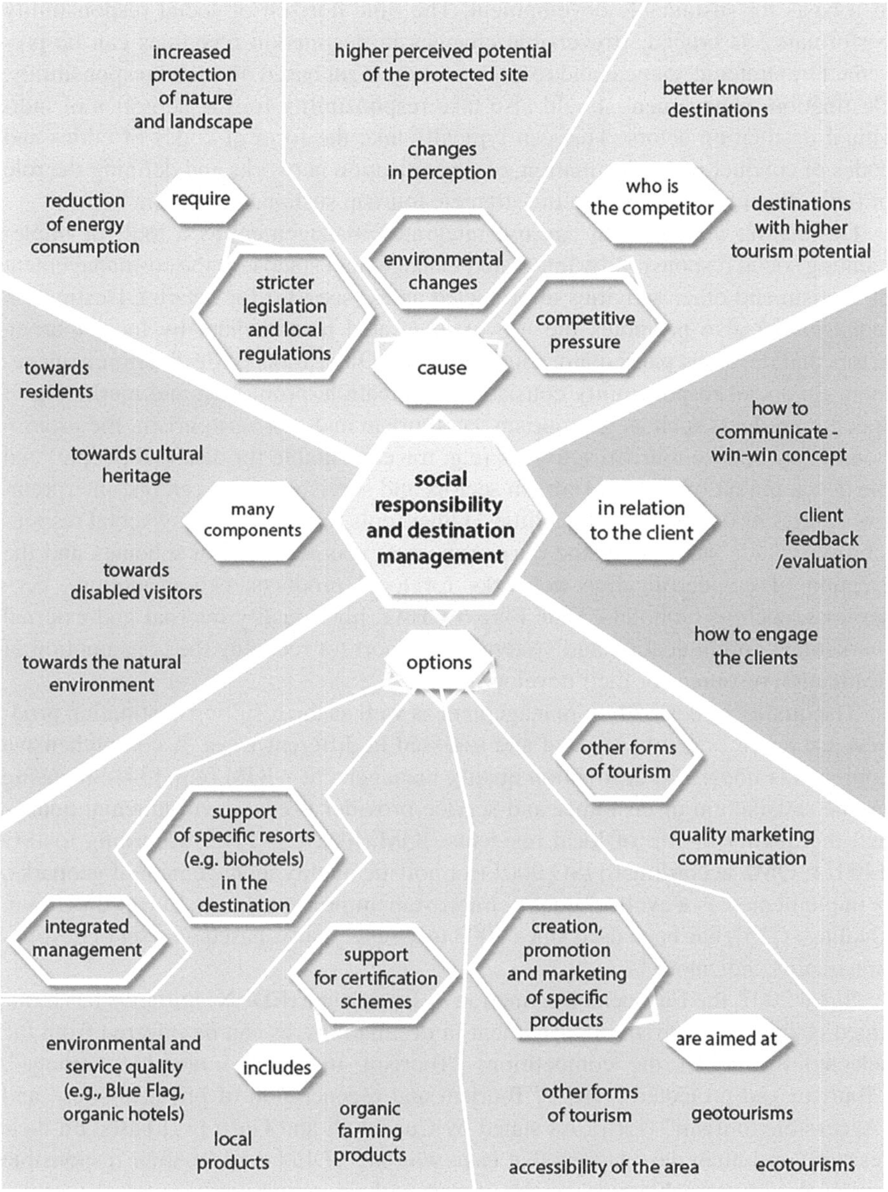

Fig. 5.5 Mind map of social responsibility—destination management

environmental and human rights legislation. Therefore, the social responsibility of destination management should be differentiated (Fig. 5.5) in relation to external factors, including competition, to destination resources in the long-term horizon of tourism development there, to tourism participants and to the development of social capital in the destination. Individual actors should perceive destination management as a basis for sustainable development. The time horizon of social responsibility performance is crucial. Irreversible changes in destination resources can be prevented by strategic, tactical and reactive management based on social responsibility. Destination management should also take responsibility for the activities of individual destination actors. This can typically take the form of codes of ethics and codes of conduct in the destination, eco-certification networks and defining the role of individual tourism actors in the strategic tourism sustainability plan.

Destination management can use integrated management as a tool for implementing social responsibility. Integrated management stands for the co-management of tourism and other activities in protected areas (see [33] for details). Destination management also promotes the use of integrated management by local tourism actors and develops participatory management. Other tools of destination management for social responsibility consist of the creation, promotion and marketing of specific products such as geotourism, ecotourism and eco-agritourism, the promotion of appropriate tourism activities (e.g. travel available for disabled people) and the development of specific tourism sectors and services as eco-centres, interpretative centres or demonstration facilities. Other tools for implementing social responsibility include support for the development of eco-certification schemes and the creation of eco-certification networks for local products, eco-agriculture, eco-services, such as biohotels, Blue Flag, ECEAT, high-quality internal and external marketing communication and systematic support for reducing the consumption of destination resources or their development.[3]

The quality of destination management as well as the resulting destination products and services can be created and assessed in different ways. A comprehensive approach is integrated destination quality management—IQM (e.g. [34]), focusing on the satisfaction of customer and service provider needs, environmental quality and the quality of life of local residents. IQM, like, e.g. EMS according to ISO 14001+, QMS according to ISO 9001+ or holistic quality management of geoparks, is implemented as a cyclical quality improvement process. In the European Union, Qualitest ([35]) has been used since 2003 to assess quality, based on a set of destination quality indicators.[4]

Since 2007, the European Destination of Excellence (EDEN) initiative has motivated responsible tourism management in destinations, as can be inferred from the selected themes of the competition: "Tourism and local intangible heritage", "Tourism and protected areas", "Tourism and regeneration of physical sites" and "Accessible tourism" ([36]). As stated by Cucculelli and Goffi [37] based on their research on Italian destinations that have won the EDEN contest, their responsible approach to sustainable tourism also increases their competitiveness.

[3] For example, support for the removal of non-native species, increasing the ecological stability of the area.

[4] Sixteen thematic areas, three indicators for each.

5.4 Social Responsibility of Management of Tourism Attractions

For many attractions, such as museums, visitor and interpretation centres, nature houses and scientific theme parks, the most significant part of their social responsibility is usually the way of interpreting phenomena, exhibits, history, evolution of life, intangible heritage etc. In addition, the visible efforts to reduce the ecological footprint of their premises by using an innovative architectural, energetic and technical solutions are other expressions of their social responsibility. Apart from its didactic dimension, it can improve the visitor's experience, as evidenced by Chung, Tyan and Lee [38] for the use of environmentally friendly technologies in museums.

For frequently visited natural and cultural attractions, well-thoughtout approaches to not exceeding the carrying capacity whilst preserving the place, including its authenticity and thus the visitors' experience, can be key components of social responsibility. In highly visited and iconic destinations such as Machu Picchu, Peru, it is a matter of finding a balance between an appropriate design of carrying capacity ([39]), conservation measures and economic interests.

As Milman, Okumus and Dickson [40] describe, for example well-managed theme parks and attractions can contribute economically and socially to the prosperity of local residents in a destination. On the contrary, according to them, poorly managed amusement parks can endanger the destination and its inhabitants and worsen their quality of life. Even some amusement parks, typical of their high attendance, declare and implement social responsibility. The two most important amusement parks Everland and Lotte World in Korea develop a wide range of activities ([41]), aimed at the local community and other stakeholders, and environmental activities. He states that, for example Everland, as the first theme park to obtain three international certifications, includes the ISO 45001 (health and safety), ISO14001 and ISO 50001 (energy) certifications. However, he emphasises that these theme parks hardly communicate these activities with the public. On the other hand, the social responsibility activities of three theme park companies in Orlando, focused on the environment, local community, customers and employees—Walt Disney World, Universal Orlando and SeaWorld Parks—are well communicated with the public ([42]). These conclusions of partial studies are summarised by Clavé, Carlà-Uhink and Freitag [43], who list social responsibility and its related sustainability as global trends in theme parks together with digital transformation and safety/security.

5.5 Social Responsibility of Tour Guides, Animators and Interpreters

The role of tour guides, animators and interpreters in stimulating the responsibility of visitors, especially in geotourism, ecotourism or ethnotourism is very important. Therefore, these tourism actors, who interpret the attractions visited, lives of local people and programme content, should themselves assume the maximum degree of responsibility towards the destination environment. It is actually a mediating role. They mediate cultural, aesthetic and ethical values by interpreting natural laws, historical events, social and political situations ([44]). The quality of their interpretation is therefore crucial for the perception of sites and destinations ([45]), for the creation of emotional attachments to destinations and even the meaning and style of travel.

The basic principle of the interpretation of the local tour guide should always be a clearly expressed responsibility towards the local ecosystems as well as towards local flora and fauna. For example, when observing birds, the guide's interpretation should educate visitors not only about the species, but also about its nesting, feeding and reproduction habits, and explain what the species needs for its life and what disturbs and threatens it. It is very effective to use stories during interpretation ([46]) and sufficient training of guides is important ([47]). The interpretation should also discourage visitors from purchasing souvenirs made from their feathers in the context of sustainability or visitor responsibility. Tour guides, mountain leaders and expedition leaders should also be aware of their responsibilities in relation to the animals used to carry cargo or transport expedition participants, as described for mules by Cousquer and Allison [48], see also Fig. 5.6.

Fig. 5.6 Horses and mules carrying visitors' food and equipment in Tayrona NP, Colombia. Responsible visitors respect local transportation methods. Tayrona NP, Colombia. Photo: J. Zelenka, 2010

The same applies to responsibility towards inanimate nature, which is performed in professional way by geotourism guides and interpreters. The emphasis on the importance of inanimate nature for living organisms must not be missing from interpretation. For example, when visiting a cave, the guide should not only explain the processes of formation of the cave and its stalactite decoration but also that the cave serves as a natural habitat for various animals with differing tolerance to disturbance. For instance, bats lose energy when disturbed in hibernation and may subsequently die. Therefore, social responsibility in relation to flora and fauna regulates visitors, prohibiting access to some cave areas, such as most of the caves in Postojna in Slovenia and in the Moravian Karst in Czechia.

On the other hand, visitors should also be responsible towards the luggage carriers like Sherpas in the Himalayas, especially in mountainous areas. They should treat them as equal beings, with respect for their knowledge and skills and their difficult work.

5.6 Social Responsibility of Visitors

The mindset, values and related motivations, expectations, perceptions, experiences and behaviour of visitors have a fundamental impact on sustainability. According to qualitative research by Caruana et al. [49], there are significant differences between different tourism actors in their perception of responsible tourism. In line with psychographic classification ([50]) and the concept of the environmental bubble ([51]:167), tourism participants can be categorised according to the degree of responsibility they exhibit towards the inhabitants and environment of the destination visited. This is a continuum, one extreme of which is the type of visitor who ignores all opportunities to reduce their own contribution to the adverse impacts of tourism. The opposite extreme is the type of visitor seeking to recognise and pursue these opportunities. It is likely that the distribution of visitors on this continuum will be analogous to Plog's [50] psychographic continuum, having the shape of a classic Gaussian curve. Allocentrics, or explorers, are usually very sensitive to the destination, the locals and the environment and generally show a high level of respect and responsibility. Psychocentrics, who are at the other end of Plog's [50] continuum, are not interested in the real life of the destination or the negative impact of tourism on the destination. Mesocentrics, who are the majority, may be interested in sustainability of tourism, but they are only claiming responsibility rather than actually fulfilling it. The reason is that this responsible approach is more time consuming, more expensive and less comfortable. Budeanu [52] confirms this, stating that an analysis of studies published to date shows a significant discrepancy between visitors' positive attitudes towards sustainable tourism and how they purchase responsible tourism products, choose environmentally responsible modes of travel such as local public transport—see Fig. 5.7—rather than taxis or rental cars, and behave responsibly towards the local community. He, therefore, recommends raising visitor awareness and also encouraging environmental awareness.

Fig. 5.7 Camels are a typical means of transport not only in North Africa. Photo: Archive of the M'Goun UNESCO Global Geopark, Morocco, 2023

In a similar vein, Constantin, Ispas and Candrea [53] discuss the need to educate tourism participants in understanding eco-certification, Han [54] recommends the application of marketing communication and awareness raising in promoting participants' interest in environmentally responsible accommodation and food service offerings, and Chen and Peng [55] describe a method of marketing communication and awareness raising on the benefits of environmentally friendly hotel services. As Chen and Peng [56] show travel agents and travel agencies can have a significant positive influence on visitor behaviour by, e.g. recommending clients to use environmentally friendly hotels.

The frequent mismatch between visitors' declared good intentions and their actual behaviour is described by Han [54] in his model of social behaviour linking the theory of planned behaviour and moral values theory. The theory of planned behaviour (Sect. 3.2.3) is often modified to more accurately explain behaviour in a particular situation—e.g. willingness to pay for environmentally friendly hotels in Kim and Han [57]. In line with the social model of behaviour presented here, various studies[5] show that visitors' fundamental relationship with the environment and the local community plays a major role in shaping their behaviour at a destination.

Ecotourism participants' motivation for environmentally friendly behaviour is the perceived value, satisfaction derived from such behaviour, and personal involvement in activities such as saving endangered species ([59]).

Tourism participants also make their environmentally and socially responsible choices by means of their spatial behaviour, preference for particular destinations or its resources such as attractions, hotels and restaurants. They may favour facilities operated by local residents, "green" accommodation and catering facilities and eco-certified establishments in general. The choice of destination itself is particularly important. For example, in the case of a disaster-affected destination, increased visitor interest and associated tourism revenue can help its revitalisation. Another expression of visitor responsibility is the selection of destinations according to the extent to which they promote and implement the principles of sustainable development. The responsible visitor may prefer to visit a national park where only Indigenous peoples are allowed to offer guiding and other tourism services such as Canaima National Park, rather than a national park where they are more likely to be an attraction or even where their human rights are not respected. The responsible visitor does not enter certain places if visitor movement in the protected area is regulated by zoning or trail infrastructure. The same applies to protected cultural monuments, commemorative areas and sacred places.

The responsible visitor willingly respects the laws, codes, regulations, as well as natural, cultural or commemorative value of a particular site, whether formally or informally expressed. In this way, the site is preserved for future users and generations, and the visitor's behaviour does not irritate the local population. Such a visitor never enters or photographs places sacred to local and Indigenous peoples, however, tempting it might be (Fig. 5.8). Responsibility is behaviourally linked to self-discipline. The responsible visitor chooses environmentally and/or socio-culturally sustainable products, such as themed holidays,[6] excursions, programmes, souvenirs, is well informed and seeks information, and directs his or her spatial and consumption behaviour on the basis of updated and correct information, on reason and feelings. He or she is interested in the specifics of the particular destination in order to understand the local history, lifestyle, value systems, cultural and natural heritage and to be able to respect and experience it all in a purposeful and conscious way.

---

[5] For example, Kim and Weiler [58] for accountability of visitors' relationship with fossil collecting.

[6] For example, production of traditional handcrafts as ceramics.

Fig. 5.8 Visitors sometimes disregard the ban on entering the sacred *ahu* (platforms) with *moai* (statues) on Easter Island and take photos standing next to the statues. Ahu and moai on Anakena Beach, the place where legends say the first inhabitants of the island landed. Photo: J. Zelenka, 2007

On the other hand, the ultimate expression of social irresponsibility is the destruction of natural and cultural heritage, typically manifested by the removal or damage of natural objects[7] and inscriptions on rocks (Fig. 5.9), cultural monuments and trees.

Visitors are addressed by general or destination-specific codes of conduct and codes of conduct for certain forms of tourism, such as ecotourism, geotourism and community-based tourism. Visitor codes for protected areas are particularly detailed and frequently supported by explanations of why the requested behaviour is needed.

5.6.1 The UNWTO Global Code of Ethics for Tourism

The Global Code of Ethics for Tourism (GCET) ([60]) is developed by the World Committee on Tourism Ethics (WCTE), an organisation under the UNWTO. GCET outlines in six points the basic objectives, principles and approaches to the broadly conceived social responsibility of visitors in relation to the natural environment, culture, economy, human rights and their own health. The Code is developing quite

[7] Living organisms or parts of them, rocks, minerals, fossils etc.

Fig. 5.9 A typical manifestation of conscious social irresponsibility is abandoned bottles, cans and inscriptions engraved on the rocks. Tacaná volcano, Mexico-Guatemala border. Photo: J. Zelenka, 2004

dynamically, as a comparison of the 2015 and 2022 versions shows. GCET is based on the fundamental idea that "*Respecting our world and all that makes us different and unique will enhance your travel experience. Whether your trip takes you to a tropical beach, a bustling city, or a tranquil village, if you act responsibly, you will earn the respect of the host communities*" ([569). The principles of a responsible visitor according to GCET ([569) are:

1. ***"Honour Your Hosts and our Common Heritage** learn about local customs, traditions, and social conditions; experience and respect all that makes a tourism destination different and unique, from its history, architecture, religion, dress, and communication codes, to its music, art and cuisine.*
2. ***Protect Our Planet**—be a guardian of natural resources, especially forests and wetlands; respect wildlife and their natural habitats; do not purchase products made of endangered plants, animals or non-sustainable hardwood; in protected areas, access only the places open to visitors; reduce waste, single-use plastic, water and energy consumption and plan your transport to cut carbon emissions.*
3. ***Support the Local Economy**—buy locally made handcrafts and products; hire local guides, spend locally and find out about tipping practices; slow down and spend more time in less visited areas.*
4. ***Travel Safely**—take health and safety precautions; practice "social distancing" during epidemics.*
5. ***Be an Informed Traveller**—choose tourism operators with environmental policies and community projects; have fun but respect the locals and their lifestyle; follow the rules at religious sites and respect "private spaces"; interact with them to understand their culture.*

6. ***Use Digital Platforms Wisely***—*before embarking on your travels, browse online reviews and blogs; provide honest reviews upon your return and promote your travel experiences on social media.*"

These principles are communicated to tourism participants in different ways using concrete, easy-to-understand activities. An example is the Ethical Travel Portal [165], a website for ethical travel with guidance on how to be a responsible tourism participant. The guiding principle of this and other similar guidelines is that the visitor is a guest at the destination and should use local resources in a sustainable way, dispose of waste, respect sacred sites, local life and the local community and learn enough about local life and customs. The visitor should also consult local residents in advance about his or her activities, especially photography, drinking alcohol, smoking, giving gifts to children on the street and courtesy gestures in communication.

5.6.2 Research on Tourism Participants' Motivation to Travel Responsibly

Motivating tourism participants to travel responsibly is important. Research on this motivation is usually carried out for different forms of responsible tourism, such as ecotourism, geotourism or community-based tourism. As shown, for example by Mody et al. [61] using the example of domestic and international visitors in India, these visitors can be as a rule divided into several different groups. In India, they identified the following segments of responsible visitors, differing in their motivation as well as socio-demographic characteristics: responsible visitors, visitors seeking new knowledge and experiences and socialising visitors. This division is, according to their interpretation, consistent with Maslow's hierarchy of needs (also [62]), where responsible visitors fulfil all levels of Maslow's needs. Research by Grimwood et al. [63] shows that the responsibility of visitors to Canada's Arctic regions is based on a *leave no trace* philosophy, which is consistent with the ethics of true tramps and boaters. At the same time, the social responsibility of visitors and their perception of the impact of attendance are important factors that shape their environmental behaviour ([64]).

Increased social responsibility of visitors can also lead to negative psychological effects. As Young, Higham and Reis [65] state, awareness of mainly environmental social responsibility can lead to negative psychological effects (including stress) for frequent air travel users. The stress results from the conflict between the motivation to be responsible, the awareness of the environmental impacts of air travel, and the desire or, in the case of long-haul flights, even the necessity to make the journey by air. Juvan and Dolnicar [66] provide similar conclusions for conservationists and their holidays in Sect. 3.3.6.

Therefore, it is important to conceive of social responsibility holistically, not only in environmental dimension but also social and economic dimensions. Otherwise, humanitarian workers, volunteers, or researchers would also have a guilty conscience. The significantly increased ecological footprint of air travel can be offset for truly conscious travellers by their responsible behaviour towards the local community, leading to an increase in the cultural identity, education and quality of life of the local population in general.

References

1. Bodhanwala, S., & Bodhanwala, R. (2022). Exploring relationship between sustainability and firm performance in travel and tourism industry: A global evidence. *Social Responsibility Journal, 18*(7), 1251–1269. https://doi.org/10.1108/SRJ-09-2020-0360
2. Bae, J.-H. (2022). Developing ESG evaluation guidelines for the tourism sector: With a focus on the hotel industry. *Sustainability, 14*(24), 16474. https://doi.org/10.3390/su142416474
3. LastFrontiers. (2016). Code of ethics for travellers, [online]. Retrieved April 10, 2016, from http://www.lastfrontiers.com/about-us/responsible-tourism/last-frontiers-code-of-ethics-for-travellers
4. Deery, M., Jago, L., & Fredline, L. (2012). Rethinking social impacts of tourism research: A new research agenda. *Tourism Management, 33*(1), 64–73. https://doi.org/10.1016/j.tourman.2011.01.026
5. IHG. (2022b). IHG green engageTM system for environmental sustainability as part of IHG's corporate social responsibility, [online]. Retrieved August 21, 2022, from https://www.ihg.com/content/us/en/about/green-engage
6. Franzoni, S. (2015). Measuring the sustainability performance of the tourism sector. *Tourism Management Perspectives, 16*, 22–27. https://doi.org/10.1016/j.tmp.2015.05.007
7. Malloy, D. C., & Fennell, D. A. (1998). Codes of ethics and tourism: An exploratory content analysis. *Tourism Management, 19*(5), 453–461. https://doi.org/10.1016/S0261-5177(98)00042-9
8. Lee, K. F. (2001). Sustainable tourism destinations: The importance of cleaner production. *Journal of Cleaner Production, 9*(4), 313–323. https://doi.org/10.1016/S0959-6526(00)00071-8
9. Akbari, M., Nazarian, A., Foroudi, P., Seyyed Amiri, N., & Ezatabadipoor, E. (2021). How corporate social responsibility contributes to strengthening brand loyalty, hotel positioning and intention to revisit? *Current Issues in Tourism, 24*(5), 1897–1917. https://doi.org/10.1080/13683500.2020.1800601
10. Kirezli, Ö., & Kuşcu, Z. K. (2012). Exploring fair trade attitude and fair trade behavior of Turkish consumers. *Procedia – Social and Behavioral Sciences, 58*, 1316–1325. https://doi.org/10.1016/j.sbspro.2012.09.1115
11. Kucukusta, D., Mak, A., & Chan, X. (2013). Corporate social responsibility practices in four and five-star hotels: Perspectives from Hong Kong visitors. *International Journal of Hospitality Management, 34*, 19–30. https://doi.org/10.1016/j.ijhm.2013.01.010
12. Gilbert, E. W., & Litt, B. (1939). The growth of Inland and seaside health resorts in England. *The Scottish Geographical Magazine, 55*(1), 16–35. https://doi.org/10.1080/00369223908735100
13. Lovelock, B. (2008). Ethical travel decisions travel agents and human rights. *Annals of Tourism Research, 35*(2), 338–358. https://doi.org/10.1016/j.annals.2007.08.004
14. Priestley, G., & Mundet, L. (1998). The post-stagnation phase of the resort cycle. *Annals of Tourism Research, 25*(1), 85–111. https://doi.org/10.1016/S0160-7383(97)00062-5
15. Chung, N., Tyan, I., & Lee, S. J. (2019). Eco-innovative museums and visitors' perceptions of corporate social responsibility. *Sustainability, 11*(20), 5744. https://doi.org/10.3390/su11205744
16. Torres-Delgado, A., & Palomeque, F. L. (2012). The growth and spread of the concept of sustainable tourism: The contribution of institutional initiatives to tourism policy. *Tourism Management Perspectives, 4*, 1–10. https://doi.org/10.1016/j.tmp.2012.05.001
17. Aminudin, N. (2013). Corporate social responsibility and employee retention of 'green' hotels. *Procedia – Social and Behavioral Sciences, 105*, 763–771. https://doi.org/10.1016/j.sbspro.2013.11.079

18. Tsai, W. H., Chang, Y.-C., Lin, S.-J., Chen, H.-C., & Chu, P.-Y. (2014). A green approach to the weight reduction of aircraft cabins. *Journal of Air Transport Management, 40*, 65–77. https://doi.org/10.1016/j.jairtraman.2014.06.004
19. Skudiene, V., & Auruskeviciene, V. (2012). The contribution of corporate social responsibility to internal employee motivation. *Baltic Journal of Management, 7*(1), 49–67. https://doi.org/10.1108/17465261211197421
20. Benavides-Velasco, C. A., Quintana-García, C., & Marchante-Lara, M. (2014). Total quality management, corporate social responsibility and performance in the hotel industry. *International Journal of Hospitality Management, 41*, 77–87. https://doi.org/10.1016/j.ijhm.2014.05.003
21. Gao, Y. L., & Mattila, A. S. (2014). Improving consumer satisfaction in green hotels: The roles of perceived warmth, perceived competence, and CSR motive. *International Journal of Hospitality Management, 42*, 20–31. https://doi.org/10.1016/j.ijhm.2014.06.003
22. Stronza, A., & Gordillo, J. (2008). Community views of ecotourism. *Annals of Tourism Research, 35*(2), 448–468. https://doi.org/10.1016/j.annals.2008.01.002
23. Ryan, C. (1991). *Recreation tourism – A social science perspective* (p. 227). Routledge.
24. Marimon, F., Llach, J., & Bernardo, M. (2011). Comparative analysis of diffusion of the ISO 14001 standard by sector of activity. *Journal of Cleaner Production, 19*(15), 1734–1744. https://doi.org/10.1016/j.jclepro.2011.06.003
25. Okazaki, E. (2008). A community-based tourism model: Its conception and use. *Journal of Sustainable Tourism, 16*(5), 511–529. https://doi.org/10.1080/09669580802159594
26. Kang, K. H., Stein, L., Heo, C. Y., & Lee, S. (2012). Consumers' willingness to pay for green initiatives of the hotel industry. *International Journal of Hospitality Management, 31*(2), 564–572. https://doi.org/10.1016/j.ijhm.2011.08.001
27. Northcote, J., & Macbeth, J. (2006). Conceptualizing yield: Sustainable tourism management. *Annals of Tourism Research, 33*(1), 199–220. https://doi.org/10.1016/j.annals.2005.10.012
28. Shaw, G., & Williams, A. M. (2002). *Critical issues in tourism: A geographical perspective* (p. 371). Blackwell Publishers.
29. Hong Kong Airport. (2022). Hong Kong – Greenest Airport, [online]. Retrieved August 04, 2022, from https://www.hongkongairport.com/iwov-resources/html/sustainability_report/eng/SR1718/world-class-gateway/greenest-airport/
30. IATA. (2022a). Enabling voluntary carbon offsetting, [online]. Retrieved August 12, 2022, from https://www.iata.org/en/programs/environment/carbon-offset/
31. easyJet. (2022). Presentation of Corporate Social Responsibility on the easyJet, plc website in 2022, [online]. Retrieved October 21, 2022, from https://www.easyjet.com/en/sustainability
32. Eccles, R. G., & Stroehle, J. (2018). Exploring social origins in the construction of ESG Measures. https://ssrn.com/abstract=3212685 or https://doi.org/10.2139/ssrn.3212685
33. Zelenka, J. (2001). Integrated management. *E+M Ekonomie a Management, 4*(1), 11–14.
34. Randle, M., Kemperman, A., & Dolnicar, S. (2019). Making cause-related corporate social responsibility (CSR) count in holiday accommodation choice. *Tourism Management, 75*, 66–77. https://doi.org/10.1016/j.tourman.2019.05.002
35. European Commission. (2005). A manual for evaluating the quality performance of tourist destinations and services, Directorate-General for Enterprise and Industry, Publications Office, p. 60, [online]. Retrieved August 19, 2022, from https://op.europa.eu/en/publication-detail/-/publication/d6a1c97f-3080-4eed-8e22-d30766542ab6
36. EGN. (2022). The EGN Charter, [online]. Retrieved August 21, 2022, from https://www.europeangeoparks.org/?page_id=357
37. Čuček, L., Klemeš, J. J., & Kravanja, Z. (2012). A review of footprint analysis tools for monitoring impacts on sustainability. *Journal of Cleaner Production, 34*, 9–20. https://doi.org/10.1016/j.jclepro.2012.02.036
38. Ciliberti, F., Pontrandolfo, P., & Scozzi, B. (2008). Investigating corporate social responsibility in supply chains: A SME perspective. *Journal of Cleaner Production, 16*(15), 1579–1588. https://doi.org/10.1016/j.jclepro.2008.04.016

39. Konopíková, M., & Soukup, M. (2014). Anthropology of tourism: From Morocco to Papua New Guinea. *Anthropologia Integra, 5*(2), 35–47. https://doi.org/10.5817/AI2014-2-35
40. Melville, S., & Ruohonen, J. (2004). The development of a remote-download system for visitor counting. In Proceedings of the second international conference on monitoring and management of visitor flows in recreational and protected areas (pp. 38–44), June 16–20, Rovaniemi, Finland, p. 417, ISBN 951-40-1930-X, [online]. Retrieved January 20, 2024, from https://mmv.boku.ac.at/refbase/files/melville_simon_ruo-2004-the_development_of_a.pdf
41. Kaur, P., Talwar, S., Madanaguli, A., Srivastava, S., & Dhir, A. (2022). Corporate social responsibility (CSR) and hospitality sector: Charting new frontiers for restaurant businesses. *Journal of Business Research, 144*, 1234–1248. https://doi.org/10.1016/j.jbusres.2022.01.067
42. Holcomb, J., Okumus, F., & Bilgihan, A. (2010). Corporate social responsibility: What are the top three Orlando theme parks reporting? *Worldwide Hospitality and Tourism Themes, 2*(3), 316–337. https://doi.org/10.1108/17554211011052230
43. CLIA. (2010). Cruise line international association, cruise market overview. Miami, FL, [online]. Retrieved August 14, 2010, from http://www.cruising.org
44. Rahman, I., Reynolds, D., & Svaren, S. (2012). How "green" are North American hotels? An exploration of low-cost adoption practices. *International Journal of Hospitality Management, 31*(3), 720–727. https://doi.org/10.1016/j.ijhm.2011.09.008
45. Wearing, S. (2001). *Volunteer tourism: Experiences that make a difference*. CAB International.
46. Bryon, J. (2012). Tour guides as storytellers – From selling to sharing. *Scandinavian Journal of Hospitality and Tourism, 12*(1), 27–43. https://doi.org/10.1080/15022250.2012.656922
47. Chung, K. C. (2020). Green marketing orientation: Achieving sustainable development in green hotel management. *Journal of Hospitality Marketing & Management, 29*(6), 722–738. https://doi.org/10.1080/19368623.2020.1693471
48. Crang, M. (2015). Travelling ethics: Valuing harmony, habitat and heritage while consuming people and places. *Geoforum, 67*, 194–203. https://doi.org/10.1016/j.geoforum.2015.06.010
49. Caruana, R., et al. (2014). Tourists' accounts of responsible tourism. *Annals of Tourism Research, 46*, 115–129. https://doi.org/10.1016/j.annals.2014.03.006
50. Pásková, M., & Zelenka, J. (2016). Social responsibility role in tourism sustainability. In P. Jedlička (Ed.), *Proceedings of Hradec economic days, February 2nd and 3rd*, ISSN 2464-6059 (pp. 777–785).
51. Cohen, E. (1979). Rethinking the sociology of tourism. *Annals of Tourism Research, 6*(1), 18–35. https://doi.org/10.1016/0160-7383(79)90092-6
52. Budeanu, A. (2007). Sustainable tourist behaviour – A discussion of opportunities for change. *International Journal of Consumer Studies, 31*, 499–508. https://doi.org/10.1111/j.1470-6431.2007.00606.x
53. Cooper, C. (1994). The destination life cycle: An update. In A. V. Seaton (Ed.), *Tourism: The state of the art* (pp. 340–346). John Wiley & Sons.
54. Han, H. (2021). Consumer behavior and environmental sustainability in tourism and hospitality: A review of theories, concepts, and latest research. *Journal of Sustainable Tourism, 29*(7), 1021–1042. https://doi.org/10.1080/09669582.2021.1903019
55. Chen, A., & Peng, N. (2014). Recommending green hotels to travel agencies' customers. *Annals of Tourism Research, 48*, 284–289. https://doi.org/10.1016/j.annals.2014.07.003
56. Chen, F.-Y. (2013). The intention and determining factors for airline passengers' participation in carbon offset schemes. *Journal of Air Transport Management, 29*, 17–22. https://doi.org/10.1016/j.jairtraman.2013.01.001
57. Kelly, J., Haider, W., Williams, P. W., & Englund, K. (2007). Stated preferences of tourists for eco-efficient destination planning options. *Tourism Management, 28*(2), 377–390. https://doi.org/10.1016/j.tourman.2006.04.015
58. Kang, J. S., Chiang, C.-F., Huangthanapan, K., & Downing, S. (2015). Corporate social responsibility and sustainability balanced scorecard: The case study of family-owned hotels. *International Journal of Hospitality Management, 48*, 124–134. https://doi.org/10.1016/j.ijhm.2015.05.001

59. Choi, H. C., & Sirakaya, E. (2006). Sustainability indicators for managing community tourism. *Tourism Management, 27*(6), 1274–1289. https://doi.org/10.1016/j.tourman.2005.05.018
60. Walker, P. A., Greiner, R., McDonald, D., & Lyne, V. (1998). The tourism futures simulator: A systems thinking approach. *Environmental Modelling & Software, 14*(1), 59–67. https://doi.org/10.1016/S1364-8152(98)00033-4
61. Michailidou, A. V., Vlachokostas, C., Moussiopoulos, N., & Maleka, D. (2016). Life cycle thinking used for assessing the environmental impacts of tourism activity for a Greek tourism destination. *Journal of Cleaner Production, 111*(Part B), 499–510. https://doi.org/10.1016/j.jclepro.2015.09.099
62. Pásková, M. (2018). Can indigenous knowledge contribute to the sustainability management of the Aspiring Rio Coco Geopark, Nicaragua? *Geosciences (Switzerland), 8*(8), 277. https://doi.org/10.3390/geosciences8080277
63. Grove, S. J. (1996). Going green in the service sector, Social responsibility issues, implications and implementation. *European Journal of Marketing, 30*(5), 56–66. https://doi.org/10.1108/03090569610118777
64. Linsheng, Z., Buckley, R., & Ting, X. (2007). Chinese perspectives on tourism eco-certification. *Annals of Tourism Research, 34*(3), 808–811. https://doi.org/10.1016/j.annals.2007.03.003
65. Wu, T.-C., Wall, G., & Tsou, L.-Y. (2017). Serious tourists: A proposition for sustainable indigenous tourism. *Current Issues in Tourism, 20*(13), 1355–1374. https://doi.org/10.1080/13683500.2014.970143
66. Janusz, G. K., & Bajdor, P. (2013). Towards to sustainable tourism – Framework, activities and dimensions. *Procedia Economics and Finance, 6*, 523–529. https://doi.org/10.1016/S2212-5671(13)00170-6

Chapter 6
Forms and Examples of Socially, Economically and Environmentally Responsible Tourism

In this chapter, inspiring approaches to the development of tourism are presented. These are examples of socially (socio-culturally), economically[1] and environmentally responsible forms of tourism. They are endorsed as an alternative to mass tourism in Kilipiris and Zardava [1]. The growing interest of visitors in these forms of tourism is an opportunity for the development of responsible tourism, which attracts the interest of other tourism actors in their "purely commercial" use. The examples given in this chapter, therefore, include a critical discussion of how these forms of tourism are implemented in line with responsible tourism. Socially responsible implementation is contrasted with the marketing and financial exploitation of this trend.

6.1 Geotourism

In particular, tourism in areas protected for their natural and scenic values must be operated in a sustainable manner. In state-protected areas, the state takes responsibility for the sustainability of its development and sets the relevant rules for tourism actors. Elsewhere, the responsibility for the heritage of the landscape lies with local people, their local authorities and other local institutions, and civil initiatives. In this context, geoparks go the extra mile. It is a concept that works with tourism actors who are able and willing to take responsibility for a specific area. Creating a geopark from the bottom up is an expression of respect for the values of the area, an act of commitment to take responsibility for the sustainability of its development. For this purpose, the managing body of the geopark creates a network of strategic partners such as museums, universities and other educational institutions, local governments and associations of local entrepreneurs (see Fig. 6.1), with whom it cooperates in educational, awareness-raising,

[1] Long-term focus of activities and their benefits, local inputs supporting other local industries—multiplication and elimination of leakages of generated capital.

M. Pásková, J. Zelenka, *Social Responsibility in Tourism*, Tourism on the Verge,
https://doi.org/10.1007/978-3-031-61609-9_6

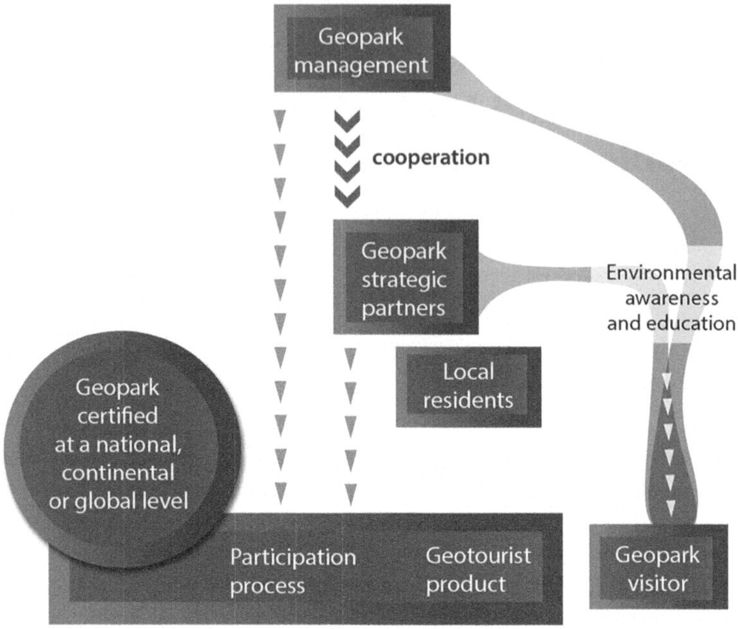

Fig. 6.1 Opportunities for participation and application of social responsibility of geopark actors

interpretation and other activities. For detailed discussions of geoparks, see, e.g. Dowling and Newsome [2], Farsani, Coelho, and Costa [3], Dowling [4] and Pásková [5].

For geoparks, social responsibility is part of their philosophy and activity profile (see Fig. 6.1). These focus on the conservation, understanding and interpretation of the Earth, its past and present geological and natural processes, and the creation of relationships with inanimate and animate nature and cultural heritages. This knowledge reflects local conditions and the lives of local people. The geopark network is developing very dynamically at national, continental and global levels ([6, 7]). Social responsibility of geoparks is enshrined in the Operational Guidelines for UNESCO Global Geoparks ([8]) and the Code of Ethics of the Global Geoparks Network ([9]). In particular, integrity, transparency, independence, impartiality, tolerance, understanding, gender equality, unacceptability of discrimination, dignity and respect for different cultures and practices are emphasised. In particular, great emphasis is placed on ensuring that geoparks and geotourism in them benefit local communities and enable developing of their culture and traditions. It also declares zero tolerance for damage to geological and palaeontological sites and for the sale of fossils and precious stones. The zero tolerance applies also to so-called rock shops, which offer frequently also stones without verified origin. They imply that their extraction can be connected with the violation of environmental laws and human rights.

Thanks to the applied concept of geoparks, geotourism is a form of sustainable tourism ([3, 4, 10]), based on learning about the evolution of the Earth through the active experience of geologically interesting landscapes with significant geological heritage ([11]). The main attractions of geotourism are the so-called geological sites, usually interpreted with the help of interpretation panels and/or QR codes and

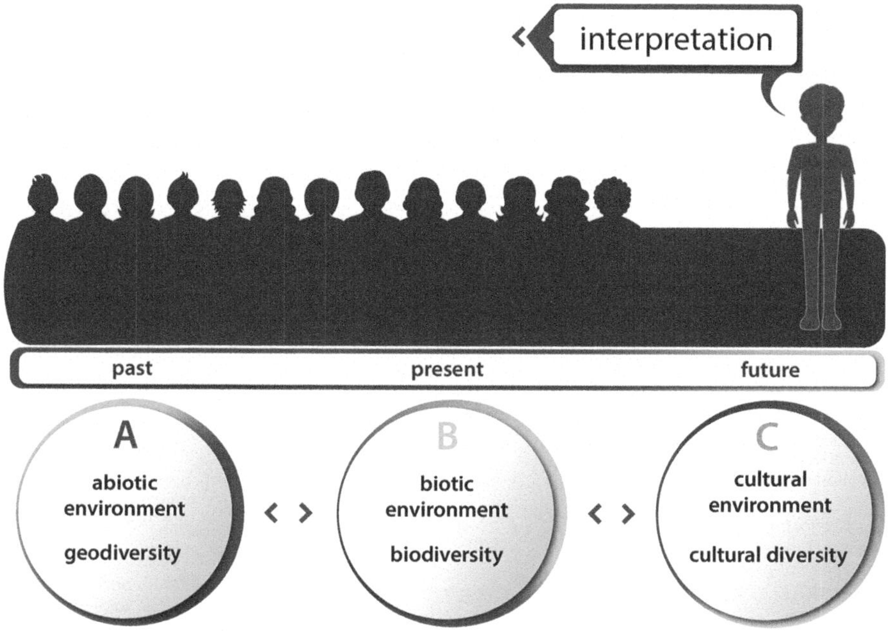

Fig. 6.2 The interpretation of the mutual connections between inanimate nature, living nature and the socio-cultural environment is the basis of geotourist interpretation. Source: Pásková [5], inspired by Dowling [4]

often linked into geotrails. Geotourism is characterised by professional interpretation and/or animation provided by a geoguide and takes place, e.g. in karst areas, rock cities, areas of palaeontological interest and former mining areas. Typical geotourist activities are visiting inactive mines, caves, amateur gold panning, georafting[2] and learning about the cultural history of the landscape.[3] Geotourism strengthens local identity, promotes local production and develops local knowledge and traditional skills through cooperation with regional academic institutions. Regional labelling of local production is used, often accompanied by stories about the origin of the product or the surrounding landscape and soil. This concept is called "GeoFood".

As Pásková [5, 12] states, geotourism has undergone an interesting evolution from a form of tourism focused on the interpretation of geological and geomorphological phenomena and processes to a holistic concept of landscape interpretation through an active experience (Figs. 6.2, 6.3 and 6.4). This holistic concept is based on the need to understand the importance of geological resources and geological tangible and intangible heritage for all spheres of human life, including science, economy, art and

[2] River rafting with interpretation of the processes of the Earth's development.

[3] Visiting archaeological sites, eco-museums, craft workshops, tasting and buying local products, accompanied by an interpretation of the origin and composition of the soil from which the products are made.

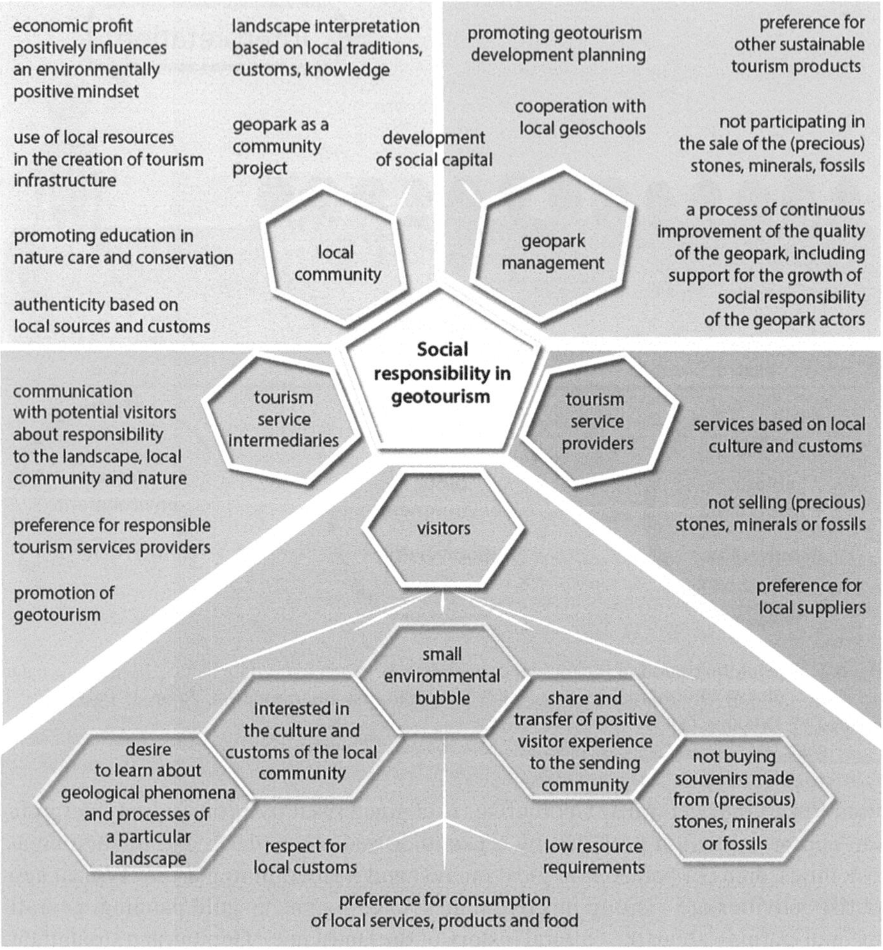

Fig. 6.3 Mind map summarising the opportunities and circumstances for the application of social responsibility by key actors in geotourism. Source: Pásková and Zelenka [13], based on findings from [2, 3, 4, 14, 15, 16, 17]

environmental protection. Intangible geological heritage refers to the knowledge of past and present geological processes. It also includes skills related to the use of geological materials and knowledge of geology in general. This also includes the prevention and reduction of geological risks, and interpretation of geological and geomorphological phenomena in the landscape as part of guiding activities.

With its holistic and systemic approach to the Earth's processes, geotourism can initiate, activate or catalyse reflections that lead to collective and individual human responsibility towards inanimate and animate nature and human society. Pásková [5] sees the search for connections between inanimate nature, living

Fig. 6.4 The Sea Factory of the Basques offers interpretation of whaling as an important cultural heritage of the Basque UNESCO Global Geopark (Spain) related to oceanic geomorphology. It is a kind of experimental archaeology based on the latest scientific knowledge. Photo: M. Pásková, 2016

nature and the socio-cultural environment as central to the application of this approach (Fig. 6.2). For this approach, Dowling [4] introduced the acronym "ABC" after the abiotic, biotic and cultural components of the environment (**A**biotic, **B**iotic, **C**ultural).

This concept of "ABC" is mainly used in the interpretation of phenomena and processes in the landscape, in which the relationship of the currently observed phenomenon is explained to the geotourist in the context of its development in the past and expected changes in the future. This holistic interpretation gives the geotourist the opportunity to understand the visited site in its complexity. Thus, he or she can appreciate the importance of minerals in the soil and groundwater for plants, the role of rock formations and caves for animal survival, and the characteristics of local rocks for their use by the local community for building dwellings, developing art, metallurgy and agriculture. This contextual understanding can significantly enhance the geotourist's understanding of his or her own responsibility towards the site they visit. It makes him or her to value the local natural and cultural heritages, to respect the primacy of local communities to decide on the management of this heritage.

The basic characteristics of the application of social responsibility in relation to the active participation of all tourism actors are presented in Fig. 6.3.

6.2 Ecotourism

The definition of ecotourism has gradually expanded, as can be demonstrated by the dynamic, widely accepted TIES (The International Ecotourism Society) definition. The original definition of ecotourism from 1990 (TIES 2015a), based on the exploration and observation of nature, included the requirement to travel in a way that does not harm the environment and improves the living conditions of local people. The 2015 TIES definition ([18]) retained the requirement for responsible travel and nature conservation and added requirements that ecotourism includes education and interpretation. The requirement to improve the living conditions of local people was changed to conservation ([18]). This definition is in line with Cobbinah [19] who further emphasises the economic benefit to the local community and visitor experience. It is also consistent with Donohoe and Needham's [20] conception, which includes conservation, education, sustainability and equitable distribution of benefits alongside nature-based tourism, as well as explicit ethical responsibility and awareness of it.

In particular, depending on the degree of social responsibility within ecotourism, a distinction has begun to be made between consumer ecotourism (also "*shallow ecotourism*" according to Acott, La Trobe and Howard [21]) and true ecotourism. True ecotourism is also referred to as real ecotourism and, according to Acott, La Trobe and Howard [21], "*deep ecotourism*". Their terminology thus corresponds to the concepts of shallow and deep ecology according to Naess [22]. Thus, consumer ecotourism can be seen as a misuse of the ecotourism "brand" because it is characterised by minimal social responsibility. In the following text, only "true" ecotourism will be discussed in the context of social responsibility. Regarding the degree and manner in which social responsibility and sustainability of ecotourism activities are practised, Weaver [23] differentiates between minimalist ecotourism and holistic ecotourism. The former emphasises megafauna and site-specific conditions. The latter is based on a globally expressed responsibility of ecotourism actors and an interpretation that includes a deep understanding of ecosystems, behavioural change of tourism actors and biodiversity conservation.

Situmorang and Mirzanti (2012) emphasise that ecotourism promotes education in nature conservation and the quality of life of the community in and around the ecotourism site. In relation to the quality of life of the local community and the optimisation of environmental, economic, psychological and political impacts on it, Scheyvens (1999) suggests consistent involvement of the local community in the planning and development of ecotourism, the highest form of which is community-based ecotourism. Coria and Calfucura (2012) consider the benefits for Indigenous people as a precondition for the successful development of ecotourism. Amongst the hindrances to its successful development, they include the lack of support for human, financial and social capital within the community, the lack of equitable ways to distribute economic benefits, and the unsafe nature of the destination. A detailed analysis of ecotourism success factors was conducted by Tsaur, Lin and Lin (2006), and based on this analysis they developed a set of ecotourism sustainability indicators.

As Liu et al. [24] show the economic benefit of ecotourism positively influences the environmentally positive mindset of local residents in community-based ecotourism. Social capital, especially its cognitive component, plays a mediating role in facilitating this process. According to Medina [25], social responsibility in ecotourism can be achieved by eco-certification, which sets rules for the participation and benefit of local residents. Responsibility in ecotourism is assured and declared mainly by means of certification schemes, codes of ethics, ecolabelling and identification with the principles of the protected area.

As shown in a study by Das and Chatterjee [26] analysing peer-reviewed articles from 2000 to 2013, and as reported by Stem et al. [27] and Stronza and Gordillo [28], inappropriately practiced ecotourism (shallow ecotourism) produces negative impacts. Ecotourism must, therefore, be based on responsibility towards nature and the local community, promoted by means of regular monitoring, educational programmes (Fig. 6.6), assessment of negative impacts and participatory destination management. The level of social responsibility, the degree of local community involvement and other aspects of sustainability in destinations with highly developed true ecotourism are summarised in Table 6.1. Such destinations are characterised by a high level of social responsibility and usually significant involvement of the local community in ecotourism activities. Ondicho [30] uses the example of Kenya's Kimana to refer to community-based ecotourism (see also Sect. 6.3). Ecotourism can take different shapes and can

Table 6.1 Level of social responsibility and level of local community involvement in selected destinations offering real ecotourism

Destination	Source	Level of social responsibility	Level of local community involvement	Comments
Lapa Rios, Costarica	[29]	High	High	An example of successful "true" ecotourism
Kimana, Kenya	[30]	High	High	Community ecotourism
Tianmushan Nature Reserve, China	[31]	High	n/a	Monitoring the status and sustainability of ecotourism
Geopark Rio Coco, Nicaragua	[10, 16]	High	High	Ecotourism as a complementary geotourism activity, using Indigenous and local knowledge and involving local people
San Pedro, Belize	[32]	High	High	Strong support for environmental protection
Gales Point, Belize		Medium	Medium	Support for environmental protection
Maya Centre, Belize		High	Low	Strong support for environmental protection
Caye Caulker, Belize		High	Medium	Strong support for environmental protection

Source: [10, 16, 29, 31, 32, 33]

Table 6.2 Conditions, preparation and implementation processes and factors influencing community-based tourism

Conditions for community-based tourism	Process of preparation for community-based tourism	Process of implementation of community-based tourism	Factors affecting community-based tourism
• Prevailing community ownership of local resources • Prevailing community decision-making on how these resources are managed • Visitors who respect the community value system, local lifestyles and local infrastructure, and who have a relatively small environmental bubble, as shown in Fig. 6.9	• The community has a clear vision for its future development • Community-based tourism is conceived as an activity meant to support the intended development of the community and should not change its functioning, philosophy or direction • The community is aware of the potential negative impacts of tourism on the quality of life for local people and their environment	• Visitors are characterised by an above-average preference for authenticity over comfort, high levels of responsibility and a below-average environmental bubble. • The number of visitors, especially overnight ones, is smaller than the number of local residents • Visitors are encouraged to fully participate in the life of the community, but their demand on infrastructure and services should not create pressure to change that life • The community should avoid creating any secondary destination resources exclusively for visitors (second "touristic world")	• Political situation • Human rights situation • Acculturation process • Institutional environment—the legislative framework of the country or region, the functioning of state and local government, the quality of destination management in the region where the community is located • Awareness, education and training in resource areas

Detailed conditions, preparation and implementation processes and factors affecting community tourism are described in Table 6.2

Fig. 6.5 Ecological interpretation and education centres in the style of local architecture, serving also local inhabitants, are an important part of the infrastructure for true ecotourism. Mudeungsan UNESCO Global Geopark, Republic of Korea. Photo: M. Pásková, 2019

complement other forms of tourism. As described by Pásková and Dowling [10] and Pásková [16] for Rio Coco, a geotourism destination in Nicaragua, ecotourism is complementary to geotourism. It uses the traditional knowledge and skills of Indigenous and local people, and seeks to enable them to be involved as much as they wish in geotourism development and to participate substantially in the related decision-making processes. Gray and Campbell [34] investigated in Costa Rica the extent to which volunteer ecotourism is an ideal form of responsible tourism. This was confirmed, with some differences of opinion, by all tourism stakeholder groups interviewed (Fig. 6.5).

The issue of social responsibility in ecotourism in relation to the involvement of all the interested actors and their respective activities is captured in the mind map in Fig. 6.7.

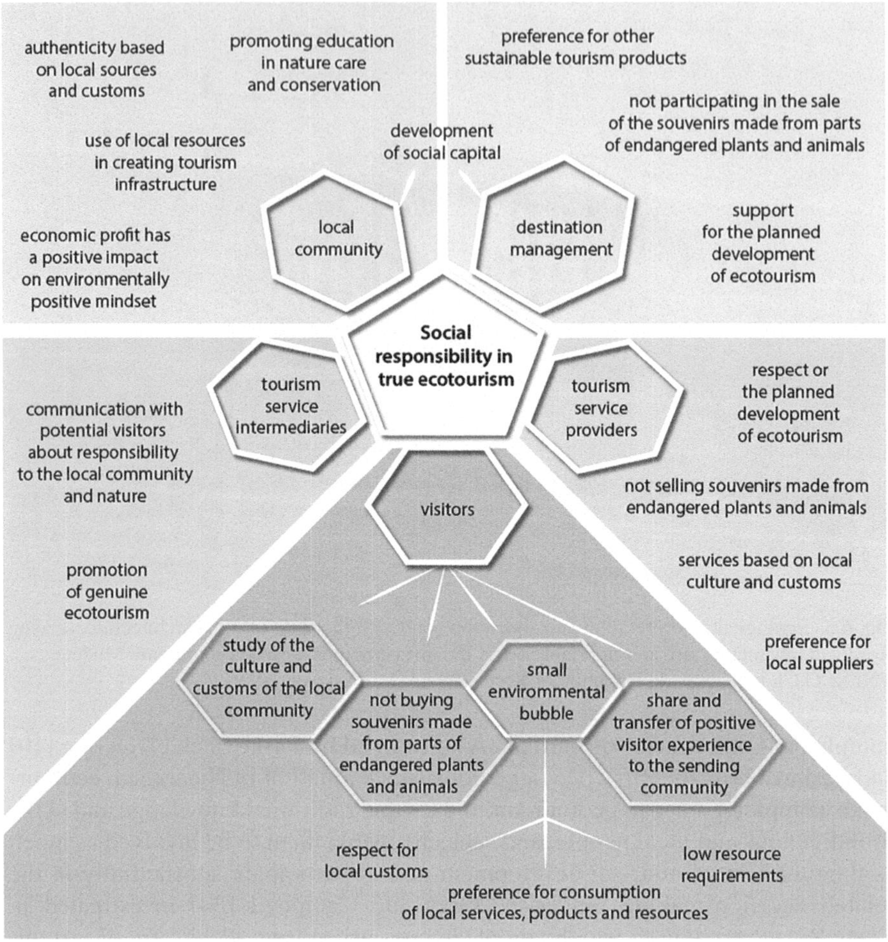

Fig. 6.6 Mind map of social responsibility of ecotourism actors and their corresponding activities. Source: [35]:1212, inspired by [24, 27, 36, 37, 38]

Fig. 6.7 Significant aspects of social responsibility in community-based tourism

6.3 Community-Based Tourism

Community-based tourism is often referred to as tourism generally developed by the local community in the territory of a municipality or other administrative unit. Community-based tourism is a form of social entrepreneurship in tourism ([45, 46, 47, 48]) and can be an implementation of justice tourism ([49]). Community-based tourism fully reflects the ideas of Higgins-Desbiolles et al. [50] on priority rights of local residents in relation to tourism.

The key here is the concept of community. It is not simply a society inhabiting a particular territorial unit. Such a society can be called a community to the extent that its members identify with it (its mission, goals and values), and to the extent that they co-own its resources and co-determine how these resources are to be used. In many countries, particularly in the Euro-Atlantic region, it is challenging to equate a municipality with a community. This is due to the growing presence of the private sector in owning local resources. Additionally, the level of influence that local residents have on the use of these municipal resources is often quite limited. Only a few municipalities have introduced what is known as a "participatory budget", and even amongst them, only a small portion of their budget is allocated to this approach. Therefore, community-based tourism is typically applied by Indigenous peoples, who often perceive natural resources as unownable and some of them even as sacred. This is their intuitive and inherited approach to understanding the concept of ecosystem services ("Mother Earth services") and a fundamental expression of their collectively established and individually carried responsibility ([10, 16]). Community-based tourism is intended and developed as an activity meant to help realise a shared vision for the further development of the community, with an emphasis on approaching the effects of globalisation. In the development and management of this type of tourism, local and Indigenous knowledge, customs, and practices are therefore prioritised ([10, 16, 51]). Conventional modern science and its knowledge systems serve primarily as sources of inspiration and guidance. Natural communities often possess distinct knowledge systems and traditions that are passed down through generations and verified over time ([10, 51, 52]). This includes knowledge of natural medicine, local cuisine, art, cosmological approaches to agriculture and caring for natural resources. Given the reasons mentioned above, community-based tourism is typically associated with a significant level of responsibility, which tends to be linked to a community's identification with its territory ([16, 53]).

Many authors, such as Okazaki[39], Zelenka et al. [54] and Pásková [53], emphasise the crucial role of involving the local community in all stages of tourism development under participatory management. This is seen as vital for ensuring the sustainability of tourism, whether it is managed by a destination organisation or other tourism entities. When the local community takes the initiative to develop tourism, especially in the initial phases of the destination life cycle as described by Butler [55] and analysed by Pásková [53], the resulting community-based tourism tends to demonstrate a high level of social responsibility. This kind of tourism, as depicted in Fig. 6.7, is well-positioned to become a "consistent" form of sustainable tourism, as characterised in Kilipiris and Zardava [1].

A study carried out in Botswana ([56]) demonstrates that strong local community participation in tourism development is a positive foundation for safeguarding local natural resources and enhancing the well-being of the local community. Consequently, community-based ecotourism can strengthen the local community's connection with conserving biodiversity and serve as a viable option or addition to other biodiversity conservation approaches, as detailed by Kiss [57]. Jones [58] emphasised the significant role of social capital in community-based ecotourism but also highlighted the potential risk it faces from ecotourism development.

In developed countries, as Joppe [59], for example argues, tourism is still developed by external tourism actors and does not derive from the activity of local communities. Community-based tourism is mainly developed in countries where the community occupies a high position in the value system, typically in the Latin American region. At the community level, the concept of community and responsibility for natural resources play a vital role in the context of community-based tourism amongst Indigenous peoples. This communal responsibility, which extends beyond tourism, is rooted in the deep connection and respect that local people have for nature and the Earth as a whole, reflecting the concept of Mother Earth. They usually reject the possibility of owning parts of it.

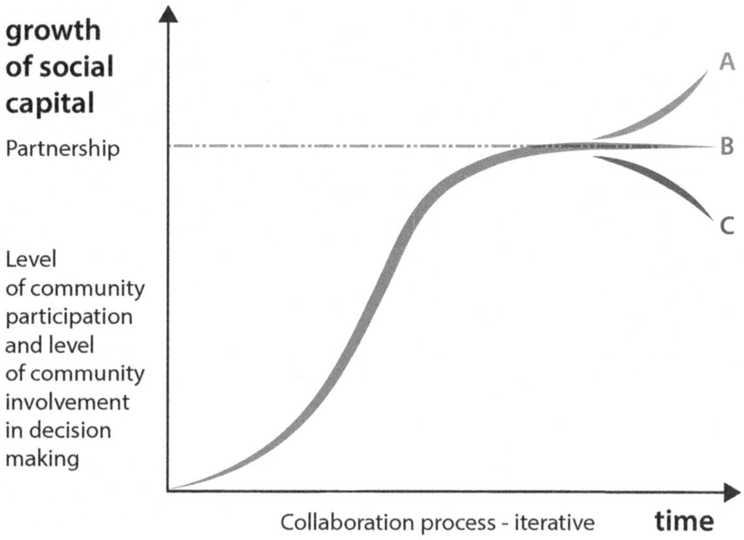

Fig. 6.8 The development of social capital reflecting the changing level of cooperation between destination actors. Source: Okazaki [39] based on Arnstein [40] and Selin and Chavez [41], adapted

Choi and Sirakaya [60] presented a comprehensive approach to monitoring the quality of community-based tourism development using sustainability indicators, which includes a list of sustainability indicators in several areas, such as human resources, policy and regulation, planning and education and financing.

Okazaki [39] presented a model of the role of social capital in community-based tourism, describing, based on the work of Arnstein [40] and Selin and Chavez [41], the process of social capital change over time as tourism actors developed their mutual cooperation. In an analogy with the destination life cycle concept ([55]), it describes the evolution of social capital using an S-curve including alternative scenarios of developments as shown in Fig. 6.8. After reaching a certain level of social capital called "partnership", their model works with three different scenarios of development of social capital during the implementation of the next stages of cooperation. Curve A describes situation in which stakeholders agree with further community participation, which leads to rapid growth with the achievement of delegated power and subsequently citizen control. Curve B depicts the rather steady state, which reflects satisfaction of stakeholders with the current situation. The last scenario described by curve C illustrates the situation when other stakeholders reject the power shift to the community, which decreases the level of social capital. The development of social capital is accompanied by transaction costs (time-consuming facilitation, conflict management etc.), which according to scenario C can even exceed the economic and other benefits of cooperation.

Jamal and Getz [61], in their analysis of the relationship between the theory of cooperation and planning theory, put forward suggestions for implementing community-based tourism with an emphasis on various aspects of social responsibility, where cooperation in planning requires:

Fig. 6.9 A fundamental feature of responsible participation in community-based tourism is the visitor's close contact with the local community and respect for their way of life even in the initial phase of the destination development. Photo: Archives of the Colca y Volcanes de Andagua UNESCO Global Geopark, Peru, 2015

- Understanding the reciprocal interdependence of tourism development with the equitable distribution of limited destination resources, the potential adverse effects of tourism on the environment and the socio-cultural surroundings, resulting in subsequent impacts on the economic yield of tourism for the local community.
- Identifying the individual and collective benefits of tourism, leading, amongst other things, to more efficient tourism development, especially through better communication and enhanced synergies, and a higher degree of environmental and socio-cultural sustainability, conflict prevention and higher resident satisfaction.
- Certainty that decisions taken will be implemented, which depends on the accountability of key actors and the availability of sufficient resources.
- The formulation of a vision, objectives, the creation of an appropriate organisation to monitor implementation and the adjustment of these objectives.

6.4 Volunteer Tourism

Volunteer tourism is a fast-growing type of travel, as noted by Mostafanezhad [62]. Its primary aim is to aid local communities by engaging participants in a variety of activities that contribute to community service during their destination visits. In a

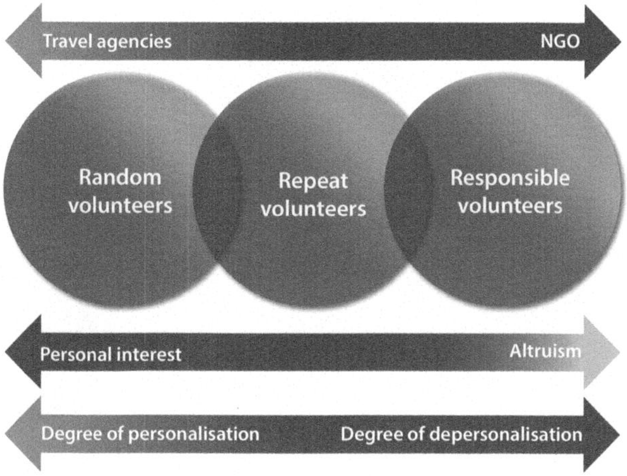

Fig. 6.10 The continuum of volunteer tourism organisers and participants. Source: Inspired by [42, 43]

thorough examination of volunteer tourism, Wearing and McGehee [63] highlighted that this form of tourism has experienced rapid growth since 1990. The focus of volunteer tourism predominantly revolves around addressing social and environmental issues in developing countries and providing assistance during and after natural disasters. Wearing [63], cit. in Wearing and McGehee [63], describes the usual activities in volunteer tourism. These activities often involve helping with conservation projects, scientific research related to wildlife, land and water, providing supportive health services, economic and social development, typically agriculture, construction, and education and restoration of heritage sites.

As Fig. 6.10 shows, it is helpful to think of the people involved in volunteer tourism on a scale, which helps understand the different aspects of this phenomenon. On one end of the scale for organisers, there are commercial tour operators, and on the other end, there are altruistic Non-Governmental Organisations (NGOs). When it comes to the participants in volunteer tourism, they fall along a continuum based on their motivations. At one end are casual volunteers who are primarily looking to satisfy their personal interests and desires, such as exploring the country or seeking new and adventurous experiences. In the middle of the continuum are recurring volunteers and at the other end are responsible volunteers. These responsible volunteers are driven by a genuine sense of altruism and the desire to help others. Altruistic or "true" volunteer tourism thus includes an altruistic volunteer, volunteer organisation and local community [63].

Volunteer tourism participants should feel a strong social responsibility towards the local community, especially in developing countries, as they significantly impact directly or indirectly the "community capitals framework" [64]. It has been defined by Flora [44] and Zahra and McGehee [64] and its components are illustrated in Fig. 6.11. Volunteering is an active expression of responsibility towards society, which is predominantly not a direct prevention or compensation of the impact of the volunteer's activities, but an expression of responsibility for others. In the case of

Fig. 6.11 Community capitals framework. Source: [29, 44]

altruistic volunteer tourism,[4] responsibility is a crucial motivating factor for participation. The volunteer tourism participant not only helps to solve problems occurring in the visited destination through his/her activity but also usually accepts limited comfort, hygiene and safety compared to standard forms of tourism. Thus, his/her environmental bubble is relatively small and generates no, or only minimal, impulses for touristification processes compared to conventional forms of tourism. In addition to altruism, other motivational factors play a role for ***commercial volunteer tourism*** participants compared to "real" volunteers.[5] These include the desire for adventure, uniqueness of experience, self-fulfilment [65], meeting other people, learning about other cultures and improving qualifications, including language skills. This is aptly characterised in Fig. 6.10, which makes it clear that the line between a commercial volunteer tourism participant and an altruistic volunteer is not a sharp one, that it is in fact a continuum of motivation and participation. The commercial volunteer is usually a so-called "random volunteer".

In any case, this form of tourism is very demanding in terms of the knowledge of the volunteer tourism participants about the local community, its values and

[4] For example, Gray, Campbell [34]—volunteer ecotourism in Costa Rica.

[5] See Sin [65] for a critical analysis of other motivational factors besides altruism.

lifestyle, both before the trip and during the stay. This is a prerequisite for the altruistic intention of the volunteer to be fulfilled in a responsible way.

Both Zahra and McGehee [64] and Sin [66] found that volunteer tourism affects the local community, much like other social, cultural and economic impacts of tourism (see also [12, 53]). These impacts vary amongst different groups and individuals within the community and are not always seen positively by everyone. In general, volunteer tourism tends to have a positive impact. This is particularly evident in the development of the social capital of the local community. Volunteers contribute to the well-being of both local people and their environment, such as managing waste, preserving biodiversity and fostering a stronger connection between living and inanimate nature.

McGehee [67] and McGehee, Santos [68] have documented the positive impact of volunteer tourism on its participants, particularly in developing their social ties, activating them and deepening their knowledge and emotions.

6.5 Slow Tourism

Slow tourism is a trend responding to a fast lifestyle, focused on performance rather than processes, procedures and consistent and long-term fulfilment of a life or corporate mission [69]. Amongst other things, this fast lifestyle leads to flattened experiences. In response to fast food and drive-in chains, slow food restaurants are emerging (e.g. [69, 70, 71, 72]). They develop the trend of experiential gastronomy, gourmet tourism, agrotourism, eco-agrotourism, ethno-agrotourism, amongst others. Likewise, there are travel companies that prioritise slow travel and the authenticity of experiences over the number of destinations visited. Such an approach to travel is founded on a holistic appreciation for the ways of life in local communities. This includes not only the local infrastructure, cuisine, etiquette, art and sports but also their pace of life, sense of time and core values. Examples include the respect for the inactivity of authorities, shops and the public during the *siesta* or frequent *fiestas* in the Iberian Peninsula, tardiness in Latin America ("mañana" concept), and the generally unhurried pace of officials and customs officials in Africa. All of this assumes a relatively small size of the environmental bubble of slow travel participants as shown in Fig. 4.4. This corresponds to Plog's [73] categorisation of participants as "allocentrics", and it also involves respecting the local people's sense of time. The size of the environmental bubble of slow travel participants is negatively correlated with the size of their level of responsibility towards the destination, the local community and its environment. Similar findings for the psychographic evolution of visitation during the destination life cycle can be found in Sect. 4.2.1.

Georgica [74] suggests that the essence of slow travel lies in the intrinsic motivation of the visitor, centred around the concept of well-being. This well-being encompasses elements such as wholesome food, physical activity, relaxation and spiritual rejuvenation. As a result, this approach, according to Georgica, materialises in the form of experiential gastronomy, where travellers actively seek out local delicacies and take the time to savour their meals in a tranquil setting, known as "slow food". They also aim to explore serene and scenic landscapes, immersing themselves deeply in the experience. This immersive experience is described by Zelenka et al. [75], involving the appreciation of the scent of plants, the sounds of the environment and the picturesque ambiance of typical locations like marketplaces.

The trend of "slowing down" tourism holds substantial potential in the realm of social responsibility. The contemporary way of life, including tourism, within Euro-Atlantic society, imposes an ever-growing strain, particularly concerning health, quality of life and the environment. This becomes evident when compared to societies with alternative, less frenetic, less consumer driven and less materialistic lifestyles. Engaging in slow tourism, which allows for a genuine and unfiltered experience, has the potential to inspire the discovery of a more natural pace of life. This can prompt reassessing life's priorities, respecting the rhythm of natural processes, and a more mindful approach to addressing current challenges without deferring them to the future. Therefore, in addition to enhancing the social responsibility of those involved in the tourism industry and consequently promoting its sustainability, the act of "slowing down" in tourism also contributes to a heightened sense of responsibility towards life in its broader sense, ultimately enhancing the sustainability of development.

References

1. Kalafatis, S. P., et al. (1999). Green marketing and Ajzen's theory of planned behaviour: A cross market examination. *Journal of Consumer Marketing, 16*(5), 441–460. https://doi.org/10.1108/07363769910289550
2. Doxey, G. V. A. (1975). Causation theory of visitor-resident irritants; methodology and research inference. The Travel Research Association Conference No. 6. San Diego: TTRA, pp. 195–198.
3. Fernandez-Feijoo, B., Romero, S., & Ruiz, S. (2014). Commitment to corporate social responsibility measured through global reporting initiative reporting: Factors affecting the behavior of companies. *Journal of Cleaner Production, 81*, 244–254. https://doi.org/10.1016/j.jclepro.2014.06.034
4. Dowling, R. K., & Newsome, D. (2006). *Geotourism* (p. 260). Butterworth-Heinemann. ISBN 9780750662154.
5. Parsa, H. G., Lord, K. R., Putrevu, S., & Kreeger, J. (2015). Corporate social and environmental responsibility in services: Will consumers pay for it? *Journal of Retailing and Consumer Services, 22*, 250–260. https://doi.org/10.1016/j.jretconser.2014.08.006
6. Parasnis, M., et al. (2016). Cleaner production in hotels and resorts. A manual for cleaner production in hotels, p. 66, [online]. Retrieved January 14, 2024, from https://www.ucalgary.ca/ev/designresearch/projects/2000/cuc/tp/outreach/cleaner%20production%20for%20hotels%20manual.pdf
7. Pásková, M. (2013). Aspiring Geoparks in the Czech National Geoparks Network. In Proceedings of the 12th European Geoparks Conference National Park of Cilento, Vallo di Diano e Alburni – Geopark- Italy, 4–7 September, pp. 201–203. ISBN 978-88-907281-4-3.
8. Tribe, J. (2009). *Philosophical issues in tourism* (p. 272). Channel View Publications. ISBN 9781845412494.

9. GH. (2022). Responsible hospitality for a better world, Green Hotelier web, [online]. Retrieved August 21, 2022, from https://sustainablehospitalityalliance.org/resources/
10. Pásková, M. (2012). Environmentalistika cestovního ruchu [Environmental tourism]. *Czech Journal of Tourism, 1*(2), 77–113.
11. Bujdosó, Z., Dávid, L., Wéber, Z., & Tenk, A. (2015). Utilization of geoheritage in tourism development. *Procedia – Social and Behavioral Sciences, 188*, 316–324. https://doi.org/10.1016/j.sbspro.2015.03.400
12. Palmer, T., & Riera, A. (2003). Tourism and environmental taxes. With special reference to the "Balearic ecotax". *Tourism Management, 24*(6), 665–674. https://doi.org/10.1016/S0261-5177(03)00046-3
13. Pásková, M. (2015). The potential of indigenous knowledge for Rio Coco Geopark Geotourism. *Elsevier B.V., Procedia Earth and Planetary Science, 15*, 886–891. https://doi.org/10.1016/j.proeps.2015.08.141
14. EGN. (2023). Global geoparks network charter, [online]. Retrieved June 11, 2023, from https://www.europeangeoparks.org/?page_id=357
15. Egri, C. P., & Ralston, D. A. (2008). Corporate responsibility: A review of international management research from 1998 to 2007. *Journal of International Management, 14*(4), 319–339. https://doi.org/10.1016/j.intman.2007.09.003
16. Parsa, H. G., & Narapareddy, V. (2015). *Sustainability, social responsibility, and innovations in the hospitality industry* (p. 370). Apple Academic Press. ISBN 9781926895673.
17. Pásková, M. (2014a). Geoturismus v geoparcích v ČR a ve světě [Geotourism in the Czech and world geoparks]. In D. Milošová et al. (Eds.), *Modul průvodce ekoturismu a geoturismu [Module for ecotourism and geotourism guides]*. Gaudeamus Hradec Králové. ISBN 978-80-7435-330-7, 158 s.
18. Tarí, J. J., Claver-Cortés, E., Pereira-Moliner, J., & Molina-Azorín, J. F. (2010). Levels of quality and environmental management in the hotel industry: Their joint influence on firm performance. *International Journal of Hospitality Management, 29*(3), 500–510. https://doi.org/10.1016/j.ijhm.2009.10.029
19. Cochran, P. L. (2007). The evolution of corporate social responsibility. *Business Horizons, 50*(6), 449–454. https://doi.org/10.1016/j.bushor.2007.06.004
20. Dowd, K., & Burke, K. J. (2013). The influence of ethical values and food choice motivations on intentions to purchase sustainably sourced foods. *Appetite, 69*, 137–144. https://doi.org/10.1016/j.appet.2013.05.024
21. Acott, T. G., La Trobe, H. L., & Howard, S. H. (1998). An evaluation of deep ecotourism and shallow ecotourism. *Journal of Sustainable Tourism, 6*(3), 238–253. https://doi.org/10.1080/09669589808667314
22. Moghaddam, Y., Bess, C., Demirkan, H., & Spohrer, J. (2016). T-shaped: The new breed of IT professional. Cutter Consortium, [online]. Retrieved February 05, 2019, from https://www.cutter.com/article/t-shaped-new-breed-it-professional-492976
23. Wang, Y.-F., Chen, S.-P., Lee, Y.-C., & Tsai, C.-T. (2013). Developing green management standards for restaurants: An application of green supply chain management. *International Journal of Hospitality Management, 34*, 263–273. https://doi.org/10.1016/j.ijhm.2013.04.001
24. Levy, S. E., & Park, S.-Y. (2011). An analysis of CSR activities in the lodging industry. *Journal of Hospitality and Tourism Management, 18*(1), 147–154. https://doi.org/10.1375/jhtm.18.1.147
25. Mayer, R., Ryley, T., & Gillingwater, D. (2012). Passenger perceptions of the green image associated with airlines. *Journal of Transport Geography, 22*, 179–186. https://doi.org/10.1016/j.jtrangeo.2012.01.007
26. Dawkins, J. (2004). Corporate responsibility: The communication challenge. *Journal of Communication Management, 9*(2), 108–119. https://doi.org/10.1108/13632540510621362
27. Social Accountability International. (2022). About SA8000, [online]. Retrieved August 08, 2022, from https://sa-intl.org/programs/sa8000/

28. Song, H., et al. (2012). Tourism economics research: A review and assessment. *Annals of Tourism Research, 39*(3), 1653–1682. https://doi.org/10.1016/j.annals.2012.05.023
29. Xing, Y., & Dangerfield, B. (2010). Modelling the sustainability of mass tourism in island tourist economies. *Journal of the Operational Research Society, 62*(2011), 1742–1752. https://doi.org/10.1057/jors.2010.77
30. NECSTouR. (2022). The network of European regions for competitive and sustainable tourism, [online]. Retrieved November 14, 2022, from https://necstour.eu/
31. Lee, K. H., Scott, N., & Packer, J. (2014). Habitus and food lifestyle: In-destination activity participation of slow food members. *Annals of Tourism Research, 48*, 207–220. https://doi.org/10.1016/j.annals.2014.06.009
32. Lee, Y.-K., Choi, J., Moon, B., & Babin, B. J. (2014). Codes of ethics, corporate philanthropy, and employee responses. *International Journal of Hospitality Management, 39*, 97–106. https://doi.org/10.1016/j.ijhm.2014.02.005
33. Nasution, H. N., & Mavondo, F. T. (2008). Customer value in the hotel industry: What managers believe they deliver and what customer experience. *International Journal of Hospitality Management, 27*(2), 204–213. https://doi.org/10.1016/j.ijhm.2007.02.003
34. Grayson, D., & Hodges, A. (2004). *Corporate social opportunity: 7 steps to make CSR work for your business* (p. 390). Greenleaf.
35. Zein, K., et al. (2008). Best environmental practices for the hotel industry. Sustainable Business Associates (sba), p. 56, [online]. Retrieved January 14, 2024, from https://nrcne.org/wp-content/uploads/2019/12/SBABGEHOTELLERIEENG2008.pdf
36. Romão, J., Neuts, B., Nijkamp, P., & Shikida, A. (2014). Determinants of trip choice, satisfaction and loyalty in an eco-tourism destination: A modelling study on the Shiretoko Peninsula, Japan. *Ecological Economics, 107*, 195–205. https://doi.org/10.1016/j.ecolecon.2014.07.019
37. Sinclair-Maragh, G., & Gursoy, D. (2015). Imperialism and tourism: The case of developing island countries. *Annals of Tourism Research, 50*, 143–158. https://doi.org/10.1016/j.annals.2014.12.001
38. Sucheran, R. (2021). Global impacts and trends of the COVID-19 pandemic on the cruise sector: A focus on South Africa. *African Journal of Hospitality, Tourism and Leisure., 10*(1), 22–39. https://doi.org/10.46222/ajhtl.19770720-84
39. Namkung, Y., & Jang, S. (2013). Effects of restaurant green practices on brand equity formation: Do green practices really matter? *International Journal of Hospitality Management, 33*, 85–95. https://doi.org/10.1016/j.ijhm.2012.06.006
40. Arnstein, S. R. (1969). A ladder of citizen participation. *Journal of the American Institute of Planners, 35*(4), 216–224. https://doi.org/10.1080/01944366908977225
41. Russo, A. P. (2002b). A stakeholders' approach to tourism policy in Bruges. In M. Jansen–Verbeke (Ed.), *Tourism studies in Bruges* (pp. 32–41). WES Research and Consultancy.
42. Cohen, E. (1972). Towards a sociology of international tourism. *Social Research, 39*(1), 164–182. https://www.jstor.org/stable/40970087
43. Wall, G. (2020). From carrying capacity to overtourism: A perspective article. *Tourism Review, 75*(1), 212–215. https://doi.org/10.1108/TR-08-2019-0356
44. Fons, M. V. S., Fierro, J. A. M., & Gómez y Patiño, M. (2011). Rural tourism: A sustainable alternative. *Applied Energy, 88*(2), 551–557. https://doi.org/10.1016/j.apenergy.2010.08.031
45. Aquino, R. S. (2022). Community change through tourism social entrepreneurship. *Annals of Tourism Research, 95*, 103442. https://doi.org/10.1016/j.annals.2022.103442
46. Aquino, R. S., Lück, M., & Schänzel, H. A. (2018). A conceptual framework of tourism social entrepreneurship for sustainable community development. *Journal of Hospitality and Tourism Management, 37*, 23–32. https://doi.org/10.1016/j.jhtm.2018.09.001
47. Jamal, T., & Camargo, B. A. (2014). Sustainable tourism, justice and an ethic of care: Toward the Just Destination. *Journal of Sustainable Tourism, 22*(1), 11–30. https://doi.org/10.1080/09669582.2013.786084
48. Selin, S., & Chavez, D. (1995). Developing and evolutionary tourism partnership model. *Annals of Tourism Research, 22*(4), 844–856. https://doi.org/10.1016/0160-7383(95)00017-X

49. Higgins-Desbiolles, F. (2008). Justice tourism and alternative globalisation. *Journal of Sustainable Tourism, 16*(3), 345–364. https://doi.org/10.1080/09669580802154132
50. Higgins-Desbiolles, F., Carnicelli, S., Krolikowski, C., Wijesinghe, G., & Boluk, K. (2019). Degrowing tourism: Rethinking tourism. *Journal of Sustainable Tourism, 27*(12), 1926–1944. https://doi.org/10.1080/09669582.2019.1601732
51. Pásková, M. (2022). Geopark certification as an efficient form of sustainable management of a geotourism destination. In V. Braga, A. Duarte, & C. S. Marques (Eds.), *Economics and management of geotourism. Tourism, hospitality & event management* (pp. 65–85, p. 124). Springer. https://doi.org/10.1007/978-3-030-89839-7_4
52. Butler, R., & Hinch, T. (Eds.). (2007). *Tourism and indigenous peoples* (p. 380). Butterworth-Heinemann. ISBN 9780750664462.
53. Park, S.-Y., & Lee, S. (2009). Financial rewards for social responsibility. A mixed picture for restaurant companies. *Cornell Hospitality Quarterly, 50*(2), 168–179. https://doi.org/10.1177/1938965509331814
54. Zelenka, J. (2015). *Cestovní ruch. Marketing [Tourism. Marketing]* (p. 247). Gaudeamus Hradec Králové. ISBN 978-80-7435-543-1.
55. Butler, R. W. (1980). The concept of a tourism area cycle of evolution: Implication for management of resources. *Canadian Geographer, 24*(1), 5–12. https://doi.org/10.1111/j.1541-0064.1980.tb00970.x
56. Russo, A. P. (2002a). The "vicious circle" of tourism development in heritage cities. *Annals of Tourism Research, 29*(1), 165–182. https://doi.org/10.1016/S0160-7383(01)00029-9
57. Kilipiris, F., & Zardava, S. (2012). Developing sustainable tourism in a changing environment: Issues for the tourism enterprises (Travel Agencies and Hospitality Enterprises). *Procedia – Social and Behavioral Sciences, 44*, 44–52. https://doi.org/10.1016/j.sbspro.2012.05.003
58. Jafari, J. (Ed.). (2000). *Encyclopedia of tourism* (p. 683). Routledge. ISBN 0415154057.
59. Jamal, T. B., & Getz, D. (1995). Collaboration theory and community tourism planning. *Annals of Tourism Research, 22*(1), 186–204. https://doi.org/10.1016/0160-7383(94)00067-3
60. Chou, C.-J. (2014). Hotels' environmental policies and employee personal environmental beliefs: Interactions and outcomes. *Tourism Management, 40*, 436–446. https://doi.org/10.1016/j.tourman.2013.08.001
61. Discovering ISO 26000. p. 20, [online]. Retrieved August 14, 2022, from https://www.iso.org/files/live/sites/isoorg/files/store/en/PUB100258.pdf
62. Millar, M., & Baloglu, S. (2011). Hotel guests' preferences for green guest room attributes. *Cornell Hospitality Quarterly, 52*(3), 302–311. https://doi.org/10.1177/1938965511409031
63. Wall, G. (1982). Cycles and capacity: Incipient theory or conceptual contradiction? *Tourism Management, 3*(3), 188–192. https://doi.org/10.1177/004728758302200158
64. Wyngaard, A. T., & de Lange, R. (2013). The effectiveness of implementing eco initiatives to recycle water and food waste in selected Cape Town hotels. *International Journal of Hospitality Management, 34*, 309–316. https://doi.org/10.1016/j.ijhm.2013.04.007
65. Sharma, A., Moon, J., & Strohbehn, C. (2014). Restaurant's decision to purchase local foods: Influence of value chain activities. *International Journal of Hospitality Management, 39*, 130–143. https://doi.org/10.1016/j.ijhm.2014.01.009
66. Sharpley, R. (2014). Host perceptions of tourism: A review of the research. *Tourism Management, 42*, 37–49. https://doi.org/10.1016/j.tourman.2013.10.007
67. Martínez, P., & del Bosque, I. R. (2013). CSR and customer loyalty: The roles of trust, customer identification with the company and satisfaction. *International Journal of Hospitality Management, 35*, 89–99. https://doi.org/10.1016/j.ijhm.2013.05.009
68. Martínez, P., Pérez, A., & del Bosque, I. R. (2013). Measuring corporate social responsibility in tourism: Development and validation of an efficient measurement scale in the hospitality industry. *Journal of Travel & Tourism Marketing, 30*(4), 365–385. https://doi.org/10.1080/10548408.2013.784154
69. Olivani, R., & Torazzo, A. (2015). Corporate social responsibility: A survey in European environmental certified tourism facilities, [online]. Retrieved May 14, 2015, from http://forumware.wu-wien.ac.at/archiv/1348193989.pdf

70. Haywood, K. M. (1988). Responsible and responsive tourism planning in the Community. *Tourism Management, 9*(2), 105–118. https://doi.org/10.1016/0261-5177(88)90020-9
71. Kumar, D. (2022). Economic and political uncertainties and sustainability disclosures in the tourism sector firms. *Tourism Economics, 29*(6), 1694–1699. https://doi.org/10.1177/13548166221113434
72. Kuo, N.-W., & Chen, P.-H. (2009). Quantifying energy use, carbon dioxide emission, and other environmental loads from island tourism based on a life cycle assessment approach. *Journal of Cleaner Production, 17*(15), 1324–1330. https://doi.org/10.1016/j.jclepro.2009.04.012
73. Pásková, M., & Zelenka, J. (2016). Social responsibility role in tourism sustainability. In P. Jedlička (Ed.), *Proceedings of Hradec economic days, February 2nd and 3rd*, ISSN 2464-6059 (pp. 777–785).
74. Getz, D. (1983). Capacity to absorb tourism. Concepts and implications for strategic planning. *Annals of Tourism Research, 10*(2), 239–262. https://doi.org/10.1016/0160-7383(83)90028-2
75. Young, A., & Kielkiewicz-Young, A. (2001). Sustainable supply network management. *Corporate Environmental Strategy, 8*(3), 260–268. https://doi.org/10.1016/S1066-7938(01)00122-1

Chapter 7
Summary

The principal drawback of the notion of tourism sustainability lies in its inherent ambiguity, multifaceted nature and, as Mihalic [1] states, the often sluggish pace of its implementation. The concept of social responsibility enables each group of tourism actors to acquire the knowledge, attitudes and mindsets required to engage in sustainable behaviour whilst undertaking tourism activities. It facilitates an understanding of the consequences of their behaviour, prompts them to consider their options and priorities, and encourages them to decide how and to what extent they will fulfill their responsibilities. At the same time, it is necessary to go beyond the "marketing framework" of declaring and implementing social responsibility and connect it in a planned and holistic manner with the solution of fundamental destination problems, with the needs of local residents and with the sustainable development of the destination [2]. But the structures of society, especially Euro-Atlantic values, do not sufficiently accommodate this responsible behaviour. In particular, there is a lack of understanding and application of the concept of ecosystem services, a strongly anthropocentric conception of the role of humans in the Earth's ecosystems, and a very limited ability to go beyond self-interest in drawing on the so-called "tragedy of the commons". The Euro-Atlantic civilisation has gradually created an attitude of entitlement, which automatically demands a continuous growth of its welfare. However, wise people say that it is impossible to grow on a finite planet. If the welfare is not currently growing, it is considered a failure of the relevant authorities. The success of this society is judged primarily by the results visibly leading to an increase in well-being and/or the well-being of other members of society. However, the consequences of their actions on the state of subsistence resources and living conditions for the future generations play a zero or marginal role in this assessment.

Facilitating more favourable conditions for the adoption and enactment of social responsibility amongst enlightened and dedicated individuals and groups is primarily concerned with a gradual change in basic attitudes regarding its role in the Earth's natural processes and ecosystems. Equally necessary is a shift towards a multidimensional perception of progress, which should no longer be confined to economic,

M. Pásková, J. Zelenka, *Social Responsibility in Tourism*, Tourism on the Verge,
https://doi.org/10.1007/978-3-031-61609-9_7

knowledge and technological aspects alone, but should also encompass humanistic and spiritual dimensions [3], all geographic levels [4] and "multispecies justice" [5]. Progress understood in this way includes issues such as the quality of education, training and research [6], the quality of the environment, human freedom, creativity and culture and the degree of social cohesion at local, national and global levels. In terms of the technical possibilities for realising social responsibility, it is necessary to change the relevant economic and social structures and mechanisms through which *the patterns of consumption* are shaped to a significant extent. This entails establishing suitable shopping, transportation and other consumption opportunities, including the use of tourism services. This requires adequate professional expertise amongst the workforce in the relevant sectors, setting appropriate legal and moral standards, developing alternative energy sources, communication networks and other infrastructure. Moreover, it involves ensuring the broadest accessibility to essential information relevant to consumer decision-making, including details such as product origin and the ecological footprint associated with a particular item. The political commitment of the state, regional or municipal authorities is essential to establish the above-mentioned legal, moral and technical frameworks. Equally important is the preference for and support of companies and organisations that apply social responsibility in their production, distribution and recycling practices, as well as encourage their customers to adopt responsible consumption behaviour. As in other sectors, within the realm of tourism, the extent and swiftness of embracing and practising social responsibility are undeniably pivotal for the future sustainability of its development, encompassing its economic, social and environmental influence. In tourism, the imperative for a "healing" process cannot be overstated, as it wields substantial influence over the economic, environmental and societal dynamics of foreign countries, people and often ethnicities during their most fragile phases of development and frequently within very vulnerable ecosystems or sacred sites. This intervention does not take place only in the context of the operation of tourism services themselves and the construction of related infrastructure; it extends to the associated *processes of acculturation* that are set in motion, for instance, by the *demonstration effect* or *environmental bubble*. Naturally, these acculturation processes are not positively selective. On the contrary, they tend to adopt behavioural patterns, social norms and lifestyles, frequently resulting in unforeseen adverse consequences for the society, economy and environment of the destination and its surroundings. The greater the willingness and capability of tourism stakeholders to shoulder their responsibility, the more pronounced the synergy becomes in enhancing the overall benefits of these efforts for society and its environment in terms of optimising the environmental, socio-cultural and economic effects of tourism.

Let us entertain the hope that this publication will serve as one of the numerous strides taken towards comprehending, embracing and meeting the social responsibility of all tourism stakeholders, including service providers, and managers. This, in turn, will help this exciting sector move towards the much-desired goal of sustainable development. At the same time, it can be an encouragement, inspiration, impulse and comprehensive basis for the research of suitable approaches and procedures and for the implementation of the concept of regenerative tourism.

References

1. McKercher, B., Wang, D., & Park, E. (2015). Social impacts as a function of place change. *Annals of Tourism Research, 50*, 52–66. https://doi.org/10.1016/j.annals.2014.11.002
2. Hughes, E., & Scheyvens, R. (2016). Corporate social responsibility in tourism post-2015: A development first approach. *Tourism Geographies, 18*(5), 469–482. https://doi.org/10.1080/14616688.2016.1208678
3. Plog, S. C. (1974). Why destinations rise and fall in popularity. *Cornell Hotel and Restaurant Administration Quarterly, 14*(4), 13–16.
4. Qiu, S., Jiang, J., Liu, X., Chen, M.-H., & Yuan, X. (2021). Can corporate social responsibility protect firm value during the COVID-19 pandemic? *International Journal of Hospitality Management, 93*, 102759. https://doi.org/10.1016/j.ijhm.2020.102759
5. Priestley, G., & Mundet, L. (1998). The post-stagnation phase of the resort cycle. *Annals of Tourism Research, 25*(1), 85–111. https://doi.org/10.1016/S0160-7383(97)00062-5
6. Alatawi, I. A., Ntim, C. G., Zras, A., & Elmagrhi, M. H. (2023). CSR, financial and non-financial performance in the tourism sector: A systematic literature review and future research agenda. *International Review of Financial Analysis, 89*, 102734. https://doi.org/10.1016/j.irfa.2023.102734

Annexes

Brief Explanation of Selected Terms

The presented interpretation of selected terms is based on the explanatory dictionary for tourism by Zelenka and Pásková [1]. It is updated and adapted to suit the context of this book.

Acculturation is the process of changes in culture and society resulting from prolonged or repeated contact between culturally different social groups in which one group can be characterised by a significantly lower level of technical development. In tourism, where the technically and economically more powerful culture travels, it is a process where residents adopt elements of the visitors' culture and elements of their culture are gradually displaced. It is a part of the so-called social exchange.

Carbon footprint is the overall volume of greenhouse gases discharged into the atmosphere as a result of the activities undertaken by an individual or organisation over a period of time. It can also be a completed process, for instance, an individual's plane flight across the Atlantic Ocean, converted to an equivalent amount of carbon dioxide (CO_2); similarly, *water footprint* is the total amount of water consumed.

Destination management is a systematic activity consisting in the application of a set of techniques, tools and measures in coordinated planning, leadership, communication, marketing, encompassing promotion, image and attractiveness of the destination, service offer and marketing research, decision-making and regulation of tourism in a specific destination, including visitor management. Apart from improvement of visitors' experience, the main objective of this continuous activity is to increase of quality of life of the local inhabitants.

Ecological footprint is a concept that involves the calculation of the spatial demand in resource consumption and the environmental impact, including waste generation, caused by a person or human population whilst engaged in a particular activity or daily life in general. It is expressed in global hectares (gha) and indicates how much land a person, an entire community within a certain geographical level or any human activity,

M. Pásková, J. Zelenka, *Social Responsibility in Tourism*, Tourism on the Verge,
https://doi.org/10.1007/978-3-031-61609-9

needs to meet their needs, including food, energy, waste, building materials and more. The concept aims, in particular, to compare different approaches, to evaluate the effectiveness of implemented measures, and to use this illustration for public communication. In tourism, the ecological footprint measures the ecological footprint of tourism as a whole sector, of a particular tourism destination, of a specific tourism activity, of a particular tourism facility and of an individual tourism participant.

Ecosystem services encompass the various functions performed by ecosystems, which can be broadly categorised into production, regulation and socio-cultural aspects, all of which establish the essential conditions for the sustenance of organisms, including humans. These services entail the generation of oxygen, the provision of food, energy, minerals and medicinal resources, as well as the capacity to retain and gradually release water across the landscape, thereby contributing to flood control, the generation of soil, the creation of microclimates and mitigating the effects of dry periods, amongst other vital climatic functions. Furthermore, ecosystem services also include biomass recycling, amongst other functions. Within tourism, these ecosystem services provide a range of benefits, particularly educational, economic, aesthetic, landscape, social, therapeutic and recreational advantages.

Environmental bubble, according to Cohen [2]:167, is the sum of the elements of culture and lifestyle that a visitor brings from his or her environment, thus significantly influencing the degree and form of tourism's impact on a destination, its people and environment, and most importantly, contributing or even initiating the process of its touristification.

Geopark is a designated area that showcases the geological heritage of a specific locality or region, with significance at the local, regional, national, continental and global levels. This heritage encompasses diverse elements like rock formations, karst landscapes, archaeological and palaeontological sites, volcanic phenomena and more. Geoparks are characterised by well-defined boundaries, engagement of local residents in their activities, management practices that advocate sustainable development and the presence of exceptional geological features. These features are interconnected through a network of clearly marked, publicly accessible geological, educational and hiking trails. In some areas, geological exhibitions and interpretation facilities may be installed to enhance the visitor experience.

Green management, often referred to as hotel GM, represents a comprehensive approach involving principles, technologies and methodology of management of accommodation facilities and other tourism establishments, predominantly hotels, with the primary objective of optimising their environmental impact. Key focus areas include conserving materials, energy through the use of passive energy buildings and combi ovens, reduced consumption of and resources, conserving and saving operational requirements, using alternative energy sources, waste management and reducing emissions of pollutants, including noise and light pollution. In a broader sense, green management also includes the promotion of healthy lifestyles, including dietary choices for guests, or support for the local, diversified economy, by the preference of local suppliers.

Indicators of sustainable tourism are indicators of sustainable development that are used to monitor the status and development of tourism in terms of its sustainability. The most commonly used set of these indicators has been developed by UNWTO and takes two forms—key indicators universally applicable in all types of tourism destinations and specific indicators for specific types of tourism destinations: for small islands, protected areas, mountain areas, places with cultural tourism attractions, for coastal areas, for settlements and ecotourism destinations.

Market externalities refer to unintended or even unwanted side effects stemming from the actions of businesses or market players who do not manage or price them in a proper way. Instead, the costs or benefits of these externalities are borne by non-market players, such as local residents. These externalities can be categorised into positive and negative types based on their impact. In the current economy, many negative externalities, like environmental pollution or alterations to landscapes, are frequently overlooked, and their potential financial expression is a challenging task. The implementation of the ecosystem services concept would be helpful.

Social capital is the sum of social relationships, mutual trust, proven personal reliability, personal respect, established ways of communication and commonly shared values related to a group of people in a tourism destination or organisation. In a tourism destination, developed social capital is an important prerequisite for the successful implementation of destination management and solving the problem of the "tragedy of commons".

Tourism destination is an area (countries, regions, human settlements and other areas) characterised by a concentration of tourist attractions, developed tourism services and other tourism infrastructure. The cohesion of such a region can be also given by its historical, cultural or natural context (e.g. ethnographic regions or protected areas).

Tourism stakeholder is any person or organisation involved in, influenced by or otherwise related to tourism in a particular destination. Typical tourism stakeholders in relation to a destination are its visitors, and residents, local government, destination management companies, public administration organisations and tourism organisations operating here.

Touristification is a gradual transformation of a place wherein the fundamental functions of a community, such as residential, service-oriented, recreational and commercial activities, give way to a predominant commercial focus on tourism. In this process, the requirements of visitors take precedence, often at the expense of the day-to-day needs and lifestyles of the local residents.

References

1. Zein, K., et al. (2008). Best environmental practices for the hotel industry. Sustainable Business Associates (sba), p. 56, [online]. Retrieved January 14, 2024, from https://nrcne.org/wp-content/uploads/2019/12/SBABGEHOTELLERIEENG2008.pdf
2. Cohen, E. (1979). Rethinking the sociology of tourism. *Annals of Tourism Research, 6*(1), 18–35. https://doi.org/10.1016/0160-7383(79)90092-6

The manufacturer's authorised representative in the EU is Springer Nature Customer Service Centre GmbH, Europaplatz 3, 69115 Heidelberg, Germany. If you have any concerns regarding our products, please contact ProductSafety@springernature.com

Printed and bound by CPI Group (UK) Ltd, Croydon, CR0 4YY
01/10/2026
02225950-0001